PAINTING THE COSMOS

PAINTING THE COSMOS

How Art and Science Intersect
to Reveal the Secrets of the Universe

NIA IMARA

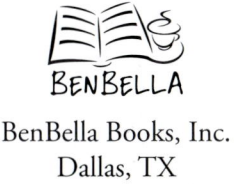

BenBella Books, Inc.
Dallas, TX

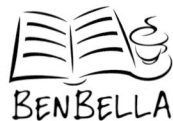

BenBella Books, Inc.
10440 N. Central Expressway
Suite 800
Dallas, TX 75231
benbellabooks.com
Send feedback to feedback@benbellabooks.com

BenBella is a federally registered trademark.

Printed in China
10 9 8 7 6 5 4 3 2 1

Library of Congress Control Number: 2024024323
ISBN 9781637742716 (paperback)
ISBN 9781637742723 (electronic)

Editing by Susanna Brougham
Copyediting by Scott Calamar
Proofreading by Becky Maines and Marissa Wold Uhrina
Indexing by WordCo Indexing Services, Inc.
Text design and composition by PerfecType, Nashville, TN
Cover design by Sarah Avinger
Cover image © Shutterstock / Fox Ave Designs
Printed by Dream Colour Printing Ltd.

For my mother,
Nehanda Imara,
who taught me to love who I come from

Contents

	Preface	ix
1.	Painting the Cosmos	3
2.	Sky Full of Stories	29
3.	Light—The Breath of Color	53
4.	Rhythm of the Sun	81
5.	Finding Life	105
6.	Space—Touching the Invisible	133
7.	Time—Light's Memory	159
8.	Harmony of the World	177
9.	Earth Quilted to Sky	197
10.	Faith	227
	Acknowledgments	253
	Image Credits	255
	Index	265

Preface

Louisiana is as far back as we remember. My father's mother was born up north to sharecropping parents in a little town called Mangham; my mother's mother, in the birthplace of jazz. They came to California when they were young, bringing with them misty recollections of *their* grandmothers. Like most black people in America, if I look back more than a few generations, the trunk and roots of my family tree seem to disperse into the haze of a plundered history. Plundered but not destroyed. I've always had a strong sense of this history—whose territory stretches well beyond Louisiana, across the Atlantic—and the culture it molded. My grandmothers were breathing vessels of that culture, incarnated in their talk, laughter, dress, cooking, work, creativity, and love.

When I was a girl I would go to church with them. Everywhere I looked, there was something to feed my hungry imagination. The dignified, friendly ushers who guided us into the pews with their perplexingly white-gloved hands; the bizarre landscape of hats; the black rainbow of faces and personalities up on the choir stand, beautiful in its variety; the passion of the pastor, with his superhuman ability to speak words that could turn people inside out; the alluring mystery of it all . . . Who was this God whom all these people had assembled to worship?

In between Sundays, I would ask my grandmothers questions: If God exists, why does He permit misery and injustice? How is it that those who have been treated so terribly can still have faith? Why does He give some the gift of belief but others not? If God directs our steps, where does free will enter in?

Many have struggled with these questions. For me, they were a matter of urgency.

My life took a decisive turn in high school, when a man with the hair and charisma of Albert Einstein stepped into my path—my first physics teacher. From the first class, it seemed to me that physics was asking the most important questions and approaching them in the most intelligent way. Physics, the bedrock of the sciences, seemed to be saying: "It is possible to excavate the hidden, fundamental workings of nature through science. Effect results from cause. Therefore, if we can discover the First Cause of everything, Its finger would point directly to God!" Through science, the meaning of God, the meaning of life, and how we should live could all be revealed.

So I imagined.

I think now that had a teacher with the voice and conviction of Martin Luther King Jr. or the grace and power of Katherine Dunham entered my life at that young age, I might have become a theologian or a dancer instead of a scientist. As fate would have it, painting would become my intractable muse, and all of these—spirituality, science, and art—would converge in my life in interesting and beautiful ways.

When people learn that I am both an artist and a scientist, they are often surprised and curious about how these two seemingly distinct worlds could collide in one individual. Where and how do the worlds of art and science intersect?

At some point, I realized that the most obvious answer lies in the reason for people's surprise: Art and science do *not* inevitably intersect. What art and science have in common is *me*—I am a crossroads at which they meet. Of course, I am not unique in this respect. I am one of countless souls within whom art and science decided to take up residence and converge.

They do not always come together naturally—not, at any rate, in today's compartmentalized consumer society. In the West, fences have been erected that cleave body from soul, play from work, spiritual life from secular life, science from art. But such fences have not always existed. For much of human history, cultures around the world practiced more organic ways of blending science and art, and both were often tied to religious faith. All three—science, art, and religion—could be integrated because all three, as philosopher Vladimir Solovyov put it, are the "spiritual offspring" of humanity.

It seems to me that the most revolutionary scientists and artists of a generation—visionaries who transform their disciplines and change the culture—are usually the most expansive. Their scope of interests and range of knowledge encompass far more territory than their particular discipline. I think of Leo Tolstoy, who considered his pedagogical writing for children his most important work and wrote a little-known series of enchanting short stories about the natural sciences. I think of John Coltrane, who took an active interest in Einstein's theories. I think of the mutual inspiration between Jasper Johns and Margaret Geller; the artist transformed the astrophysicist's pioneering map of the universe's distribution of galaxies into a work of art. And George Washington Carver—this consummate artist painted throughout his life, spoke to the flowers, and always acknowledged the leading role that faith in God played in his science.

For artists and scientists such as these, the fence does not exist. Art and science may intersect within individuals in ways as singular as their personalities.

In our fragmented society, art and science are usually kept separate (unless the artist or scientist intentionally brings them together); nevertheless, they cannot help but influence each other. As Tolstoy expressed it, "Science and art are as closely bound together as the lungs and the heart, so that if one organ is vitiated the other cannot act rightly." Today neither of these two vital organs of humanity is functioning altogether "rightly." Science is used to design weapons of mass destruction and to manufacture needless commodities whose mass production is destroying the environment. Art, meanwhile, is used to sell those needless commodities and to encourage rather than alleviate violence.

Since taking my first physics class, I have outgrown the naive belief that science holds answers to those big existential questions about life's meaning.

We impart meaning to things. Each of us decides what their life means. Experience has tempered my expectations for what science can do for us and raised my standards for what it *should* do.

Today I am drawn to science because it deepens my sense of the mystery of the universe and because the lessons we learn from nature can aid our spiritual development. Science stirs my curiosity—a doorway to an expanded, higher perspective. If the desire to know the God of the universe is what led me to science, it was the desire to know the God within that led me to art. Art puts me in direct contact with my heart and spirit and with the hearts and spirits of others. For me, science is the search from the outside in, while art is the search from the inside out. Science sharpens my intellectual sight; art helps me to see through the eyes of the heart.

The questions that my grandmothers' faith raised within me are with me still. The one thing that faith settled early on and with certainty is the reality of love. How we see the universe and our place in it can expand or shrink our capacity to love. Do we view ourselves as insignificant motes of dust in an indifferent universe? As isolated individuals who must compete to survive? Or do we recognize the interdependence and infinite preciousness of all life?

I hope this book sheds light on some of the fascinating ways in which scientists and artists understand the universe. My bigger hope is that it encourages a feeling of interconnectedness and unity—within ourselves, with one another, with the entire cosmos.

PAINTING THE COSMOS

CHAPTER ONE

PAINTING THE COSMOS

The arts and sciences are avatars of human creativity.
— *Mae Jemison* —

To be creative means to connect. It's to abolish the gap between
the body, the mind and the soul, between science and art,
between fiction and nonfiction.
— *Nawal El Saadawi* —

Nowadays, science and art are considered separate things. For most purposes, we put science over *here* and art over *there*. Art is something we expect to experience when we visit a museum, concert hall, or theater, or when we open a novel. Science, meanwhile, belongs in labs, research institutes, and observatories. Not only have art and science been relegated to different places, but they are often seen as opposed, requiring completely opposite skill sets and ways of thinking. But it was not always this way. Before the words "science" and "art" took on the distinct meanings they have today, societies around the world mixed them naturally. Art and science as we understand them now are inventions just hundreds of years old.

This separation of art and science is consistent with the many other divisions that have come to seem natural in our fractured society. Western culture is full of dichotomies: human versus nature, body versus soul, white versus black, logic versus emotion—science versus religion, spirituality, and art. We often take these divisions for granted, as if they were natural laws, but they are entirely human made. They impact our thinking and well-being—as individuals and communities—in many ways. It is not difficult to understand why people have a strong hunger for wholeness and unity. Perhaps part of the way to address this need is to rejoin what has been severed.

This book is not concerned with trying to merge art and science for the sake of it—such arbitrary marriages have been forced frequently enough—but with achieving a deeper understanding by seeking lessons from both. This little book starts with a simple question: What can we gain by looking at science and art together? By exploring them jointly, I hope to provide a small glimpse of the richer, deeper understanding and sense of coherence we can achieve once we remove art and science from their lonely silos.

Life is an unfoldment, and the further we travel the more truth
we can comprehend. To understand the things that are at our door
is the best preparation for understanding those that lie beyond.
— Hypatia —

One of my favorite examples of the marrying of art and science is the Great Pyramid at Giza in Egypt, the oldest of the Seven Wonders of the Ancient World and the only one still standing. The Great Pyramid was built 4,500 years ago for Pharaoh Khufu. The Egyptian pharaohs believed they would

WONDER OF THE WORLD | The pyramids of Giza as seen from the International Space Station.

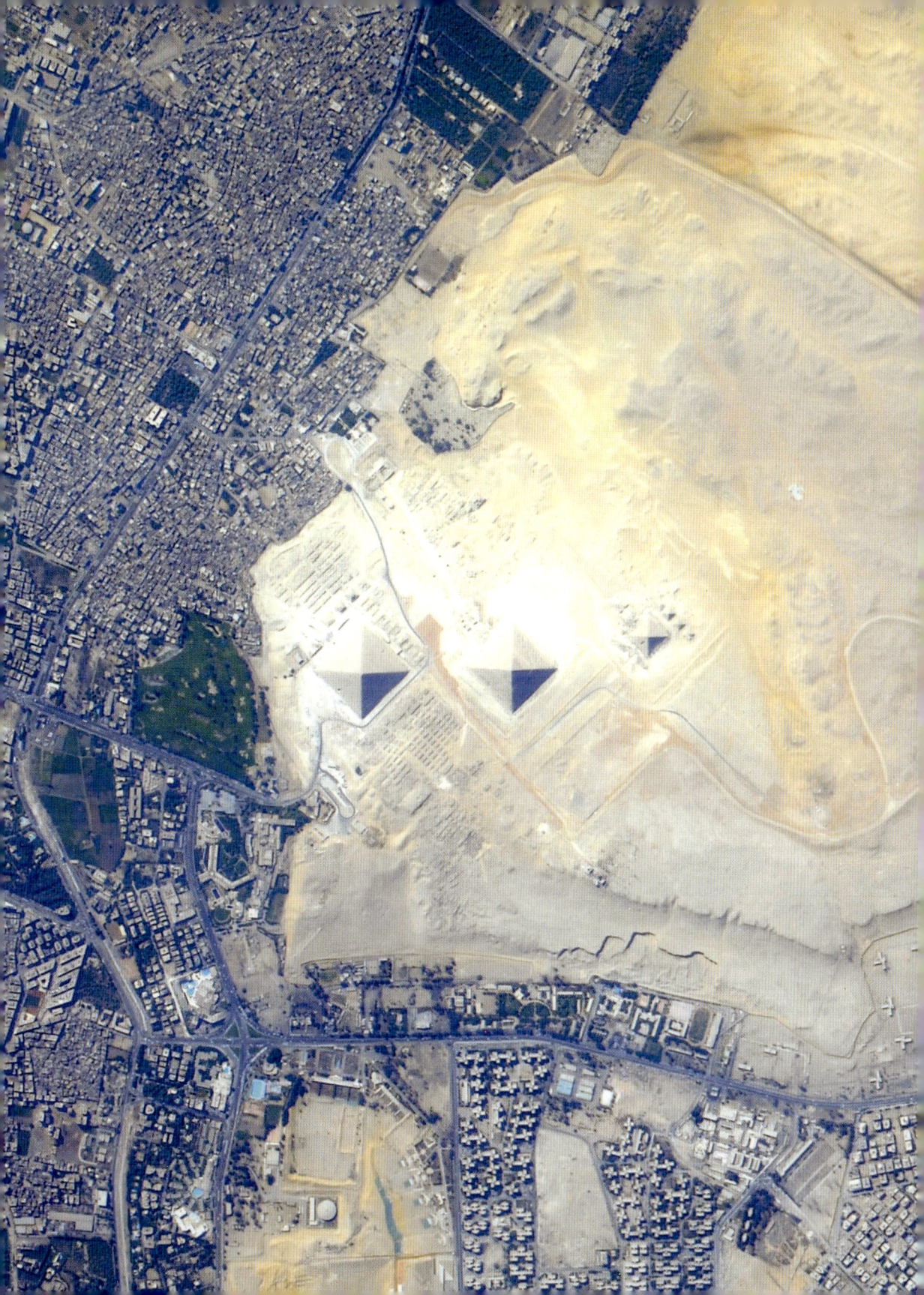

become gods after they died and had their people construct magnificent tombs to house the treasures and amenities they would need in the afterlife. Built about five miles from the Nile River, the Great Pyramid was the first and largest of the three main pyramids to be erected at Giza. The iconic site also includes three smaller pyramids and the Great Sphinx.

For inhabitants of the Libyan Desert in the 26th century BCE, the Great Pyramid must have been awe-inspiring and humbling—even more than it is for us today. Constructed from 2.3 million blocks of stone and ascending to a height of 482 feet (147 meters), the Great Pyramid would have been the biggest monument anyone had ever seen at the time. For more than 4,000 years, until the Eiffel Tower was built in the late 1880s, it held the record as the tallest building in the world. The stones were originally encased in white limestone polished to a high gleam, so that the monument sparkled like a six-million-ton jewel visible from more than 20 miles (32 kilometers) away on a clear day. Though its color gradually eroded to a relatively modest tawny yellow, the Great Pyramid still stands out as one of the few human-made structures easily visible from space!

A majestic tomb to glorify dead royalty and inspire awe in the living, Giza is both a grand work of art and a work of science. I like to think of it as a massive astronomical observatory, with every part serving a specific purpose: to connect all those who experience it—dead or alive—with the heavens.

The base of the monument is a near-perfect square, with each side stretching 756 feet—the height of a 50-story building—and aligning almost exactly with the cardinal directions: north, south, east, and west. Narrow passageways tunneling from interior chambers to the surface appear to point toward stars that may have been important to the Egyptians. One of them pointed in the direction of Thuban, the former North Star—the star in the Northern Hemisphere toward which Earth's axis points. Thuban is not, however, the North Star that we see today. As Earth rotates about its axis, it wobbles like a spinning top. Due to this slow wobble—a phenomenon called *precession*—Earth's axis does not continuously point to the same spot in the sky. Rather, it gradually comes to point at another star, then later at another, and so on. Five thousand years ago, for the ancient Egyptians, the North Star was Thuban.

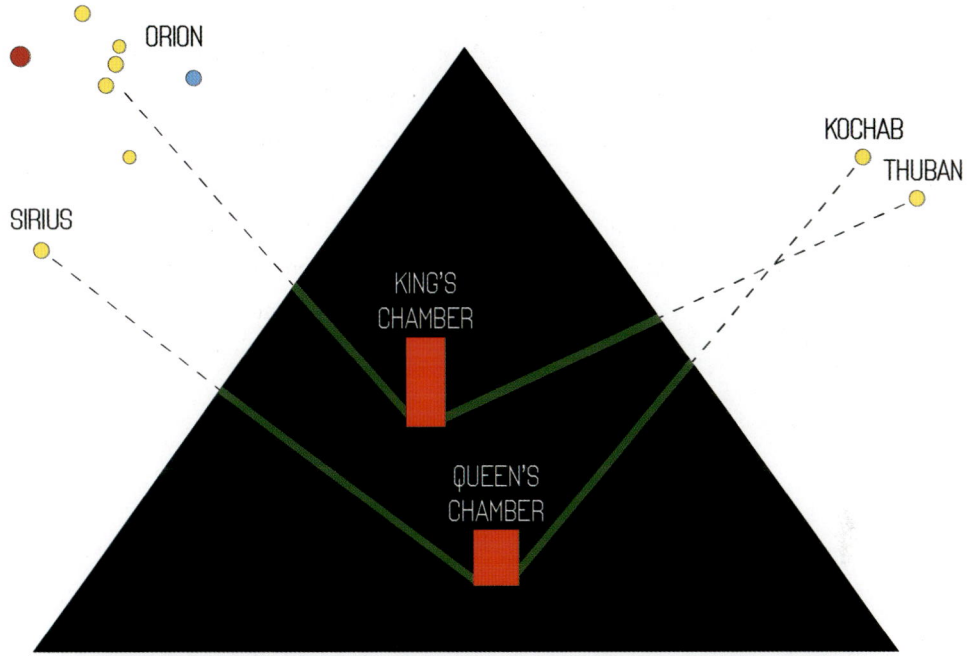

CROSS SECTION OF GIZA | Possible star alignments of the shafts in the Great Pyramid.

Today it is Polaris; by the year 4000, it will be Errai, a modest star located in the constellation of Cepheus. Some 20,000 years from now, Thuban will reclaim the title of North Star.

Scholars still debate the purposes of the shafts, the cardinal alignments, the precise geometry, and many other impressive features of the Great Pyramid. Whatever these design choices meant to the ancient Africans, it is clear that they were intentional, that every stone, surface, corner, wall, and room of the royal tomb had deep symbolic significance. The square, for instance, was for the Egyptians a most sacred symbol, and references to the shape can be found repeatedly in their architecture and art. The whole geometry of the pyramid is thought to symbolize Benben, which in ancient Egyptian cosmology was the first mound of land to emerge from the primordial waters when the world was created. Overall, the elegant form of the structure and its steeply sloping walls suggest communication with the cosmos, a sort of channel—or union—between the material and spiritual worlds.

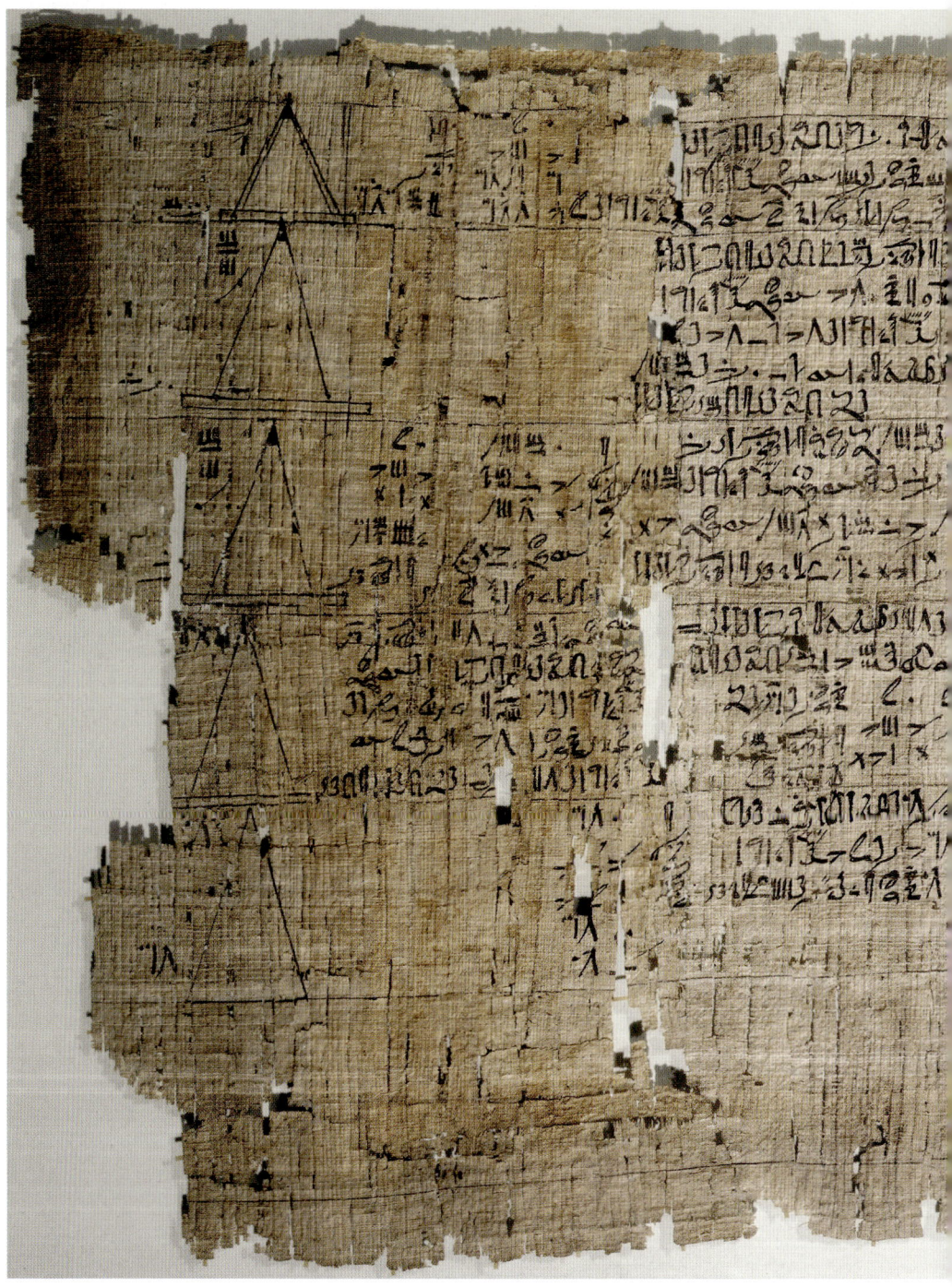

RHIND PAPYRUS | A segment of the so-called Rhind Papyrus, named after the Scottish archaeologist who bought it in 1858. Also known as the Ahmes Papyrus, in honor of the Egyptian scribe who copied it in about 1550 BCE, the document is nearly 4,000 years old, almost as old as the Great Pyramid. It contains several mathematical formulas, including one showing how to calculate the slopes of pyramids.

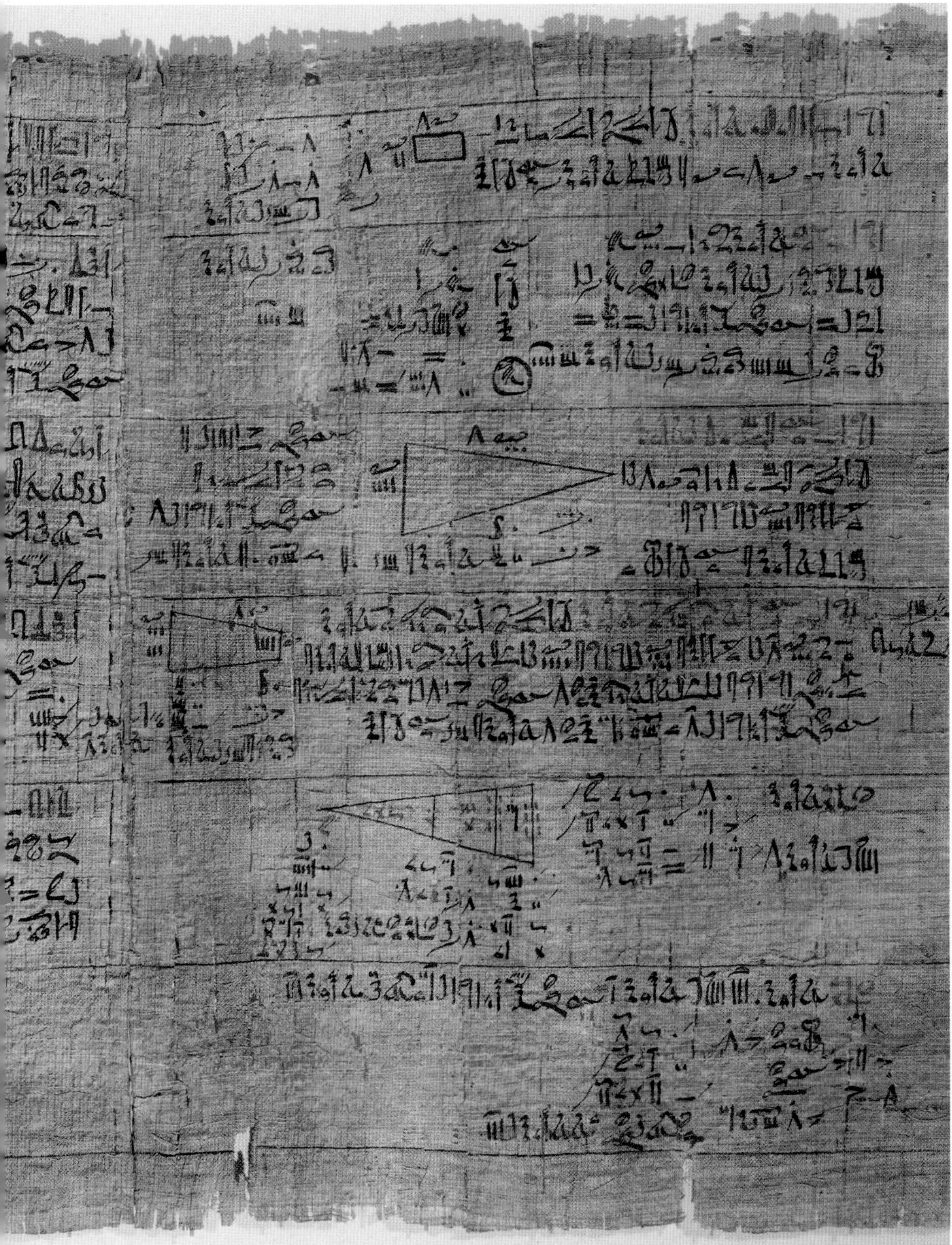

To conceive and execute such a phenomenal vision required a sophisticated understanding of physics, mathematics, engineering, and astronomy. The foundations of this system of science were laid hundreds, if not thousands, of years earlier in Africa. Scrolls of papyrus from the period, works of art in themselves, were inscribed with detailed mathematical formulas and instructions needed to pull off this extraordinary feat. The serious business of ushering kings and queens into the afterlife required both art and science, working hand in hand.

To Be Human

Art and science are essential to the human experience. They help us fulfill deeply human needs: to understand the world, to create, and to connect with others. Our earliest ancestors were motivated to understand their environment in order to survive, adapt, and find some measure of security in an often unpredictable world. We may think that today, art and science can enhance our lives, but we don't have to worry about survival to the same extent. That may not be entirely true. The environment we live in, with its ever-changing technology and hectic pace, can make it a challenge to live well and thrive. The way of art has a lot to offer us now, as it has in the past. Art is a defining characteristic of what it means to be human. The fact that humans have been engaging in this creative activity for so long testifies to its capacity to ground us and inspire us.

In 2018, archaeologists discovered the oldest human drawing yet uncovered. It was found in Blombos Cave, located on the southern Cape coast of South Africa. Scientists date the small, abstract drawing, made with ocher on the surface of an unassuming rock, at 73,000 years old. The rock displays geometric marks in a crosshatched pattern, similar to other drawings and engravings found at the site. It's unclear what the drawing meant to its creator. Was it a message? A warning signal? A charm or a talisman? An absent-minded doodle that developed into something more? Without knowing the creator's intention, how can we even qualify it as "art"? (This, of course, depends on how we define art, a question I'll come back to soon.)

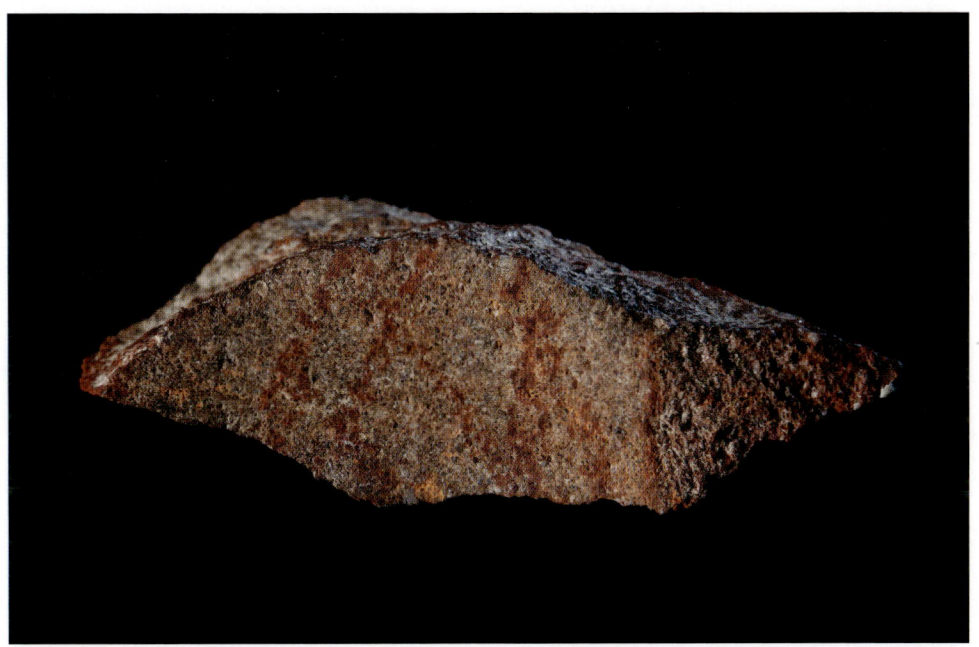

ANCIENT ART | At 73,000 years old, this rock art found in Blombos Cave is the oldest human drawing yet discovered. *IMAGE COURTESY OF PROFESSOR CHRISTOPHER HENSHILWOOD.*

One thing is clear: this creation *did* require intention. The artist, first of all, had to have an idea, a vision, or at least a strong intuitive sense of what she wanted to communicate. She then had to go out and find materials to fulfill that intention. It was probably not very difficult to find some ocher. Ranging in color from yellow to purple, this pigment occurs naturally in iron-rich rocks and soil. It's the basis for the oldest form of paint, used around the world for thousands of years to decorate walls, fabric, and the human body.

Then, for the artist, came the exacting part of the task: the execution of the drawing. She painstakingly marked a series of diagonal lines on a surface barely 1.5 inches (3.8 centimeters) in length. The quality of line, though faded, shows clear intention. If this little chunk of rock was a fragment from a much larger piece, imagine the planning, patience, and heart that went into the complete work!

Leaping forward to the end of the Paleolithic era—about 45,000 years ago—we begin to find examples of representational art possibly connected with the cosmos. Cave paintings, carvings, and figurines from Africa, Asia, and Europe reveal the main preoccupations of our Stone Age ancestors: food, fertility, and nature. Recent studies suggest that some cave paintings, located in modern-day Türkiye, Spain, and France, are not merely representations of wild animals but also signify certain constellations. It is possible that early humans had advanced knowledge of the sky and developed sophisticated systems of timekeeping based on the motions of the stars. Findings suggest that they were even aware of precession, the slow wobble of Earth's axis over the course of millennia.

Astronomy is perhaps humanity's oldest science. Long before the invention of telescopes, people could look up and observe the orderly motions of the cosmos. Motivated by the desire to understand their environment and thrive within it, our early ancestors carefully observed the cycles of the stars so they could predict the changing of seasons, determine when to plant their crops, navigate to unexplored territories, and connect with their gods. Every organized society that has existed has developed a practice of astronomy.

The Aboriginal people of what came to be known as Australia were diligent observers of the Sun, Moon, stars, and planets for thousands of years before the British colonized their lands. Wurdi Youang, an arrangement of stones built by the Wathaurong people near the southeast coast of Australia, is perhaps the oldest existing astronomy site in the world. Basalt stones form an egg-shaped enclosure roughly 164 feet (50 meters) in diameter, with the three biggest stones grouped on the western side. Scientists have found that the east-west axis of the arrangement aligns with the position of the Sun at sunset on the solstices and equinoxes to within a few degrees. Carbon dating studies at neighboring sites suggest that Wurdi Youang could be 11,000 years old.

Nabta Playa, located in southern Egypt, is another ancient astronomy observatory. The stone circle was built more than 9,000 years ago, perhaps by a cattle-herding people who settled there during the wet summer season, when the desert basin where they usually lived was likely to be flooded. The massive stones are aligned with important stars, such as Sirius, Arcturus, and those in Orion's belt. The site acts as a sort of enormous astronomical

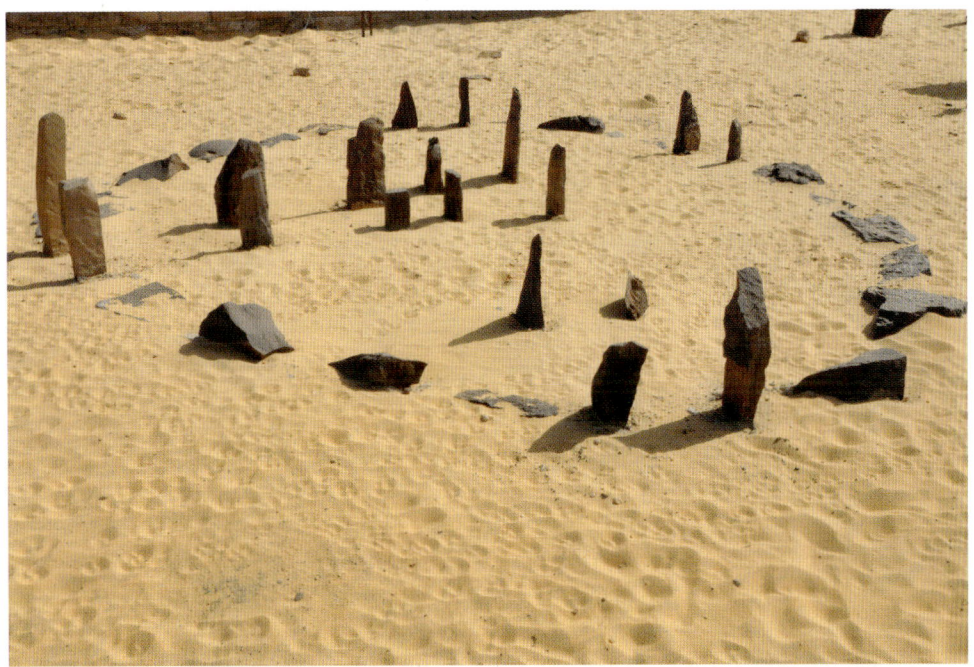

NABTA PLAYA | A reconstruction of Nabta Playa at the Nubian Museum in Aswan, Egypt.

clock to mark the seasons, guiding farmers as to when to plant and harvest. Like other stone circles erected in Africa and Europe, Nabta Playa also served as a cultural and religious site, where ceremonies were performed and the dead buried.

Indigenous people throughout pre-Columbian and pre-colonial America left behind art and artifacts showing how their relationship to the cosmos was integrated into their daily lives and provided structure for their societies. The medicine wheel or sacred hoop, a symbol with origins in various Indigenous American tribes, is one such example.* Hundreds of these structures, built right into the land, exist across North America. The medicine wheel in Bighorn, Wyoming, was created some 600 or 700 years ago, perhaps by ancestors of the Cheyenne people. Constructed of huge boulders, it has 28 spokes,

---

* The terms "medicine wheel" and "sacred hoop" were probably not coined by Indigenous peoples.

corresponding to the number of days in the lunar month. Alignments about the hub of the wheel mark the position of sunrise and sunset during the summer solstice, as well as the rising positions of three bright stars: Aldebaran, Sirius, and Fomalhaut.

Much of the knowledge developed by our early ancestors around the globe was lost or destroyed during various conquests throughout history, including colonialism. In many parts of the world, European ideas about art and science continue to dominate. And mainstream scholars have devoted far less attention to non-European art and science. Sadly, we still need to be reminded that cultures around the world and throughout history have contributed to science and art, and these fundamentally *human* endeavors belong to the heritage of the world.

Gods of Knowledge

As we talk about the art and science of earlier times and different cultures, it's worth remembering that ideas about art and science are not universal. Art and science are culturally situated, and context matters.

I recently went to an exhibit on African art at the Seattle Art Museum. In one room was a group of several dozen masks from various tribes in West and Central Africa. The masks stood in rows, like a rank of soldiers or schoolchildren lined up for a class photo. I saw masks from the Dan people of Côte d'Ivoire and Liberia, the Bwa of Burkina Faso, and the Dogon masks of Mali. They were attached to the ends of long poles, so that they all peered at you at about eye level. In the low lighting and library-quiet atmosphere of the museum, they created a stunning effect. I wondered: What was the curator hoping for visitors to take away from this display?

Even while acknowledging that Africa is a culturally diverse continent, we can state a general truth: traditional African art is made for specific contexts, often serving religious and social functions. Masks like the ones displayed in the Seattle Art Museum that day are typically created for specific purposes—such as rite-of-passage ceremonies—and may even be kept from public view until those special times of year when they are needed.

They are believed to be inhabited by spirits, often represented by animals or heavenly bodies, like the Sun. They are frequently worn with elaborate costumes. Balancing on poles there on the gallery floor, they looked like de-bodied specters.

In African cultures and many others, art has traditionally been tied to its particular context. For much of history, scientific pursuits were also deeply contextual and often motivated by religious, political, and civic concerns. Moreover, there were no hard distinctions between "art" and "craft," chemistry and alchemy, astronomy and astrology. Because these classifications did not exist, many cultures did not have words for "art" and "science" in the ways we define them today.

Our current systems of art and science in the Western world are fairly recent creations in human history. They are largely European inventions that emerged during the Enlightenment, a cultural movement that transformed Western thought during the 17th and 18th centuries and, ultimately, impacted the globe. It was during this period that the concept of *fine art* arose. Around the same time, the way Europeans thought about *science* and what belonged in that category also underwent a big transformation.

Before then, most cultures, European included, didn't consider things like music, theater, literature, sculpture, and painting as belonging to a single group called *art*. In ancient Greece, the closest equivalent to our word "art" was *techne*, meaning "technique," "skill," or "craft." *Techne* is concerned with technical skill or specialized expertise, and for the ancient Greeks, that included the fields of sculpture, poetry, carpentry, harp playing, and medicine.

The early Hellenists did not have a specific word for "science" either, though a close equivalent would be *episteme*, usually translated as "knowledge" or "scientific knowledge." *Episteme* and *techne* were closely related. Socrates, for instance, thought that many activities associated with *techne*, such as astronomy and painting, belonged to the domain of *episteme* as well. And *episteme* was not the only form of knowledge that Greek philosophers defined. There was also *doxa*, which is similar to "opinion," or the type of knowledge learned through hearsay. The deepest type of knowledge is *gnosis*, a spiritual awareness that penetrates to the true nature of things.

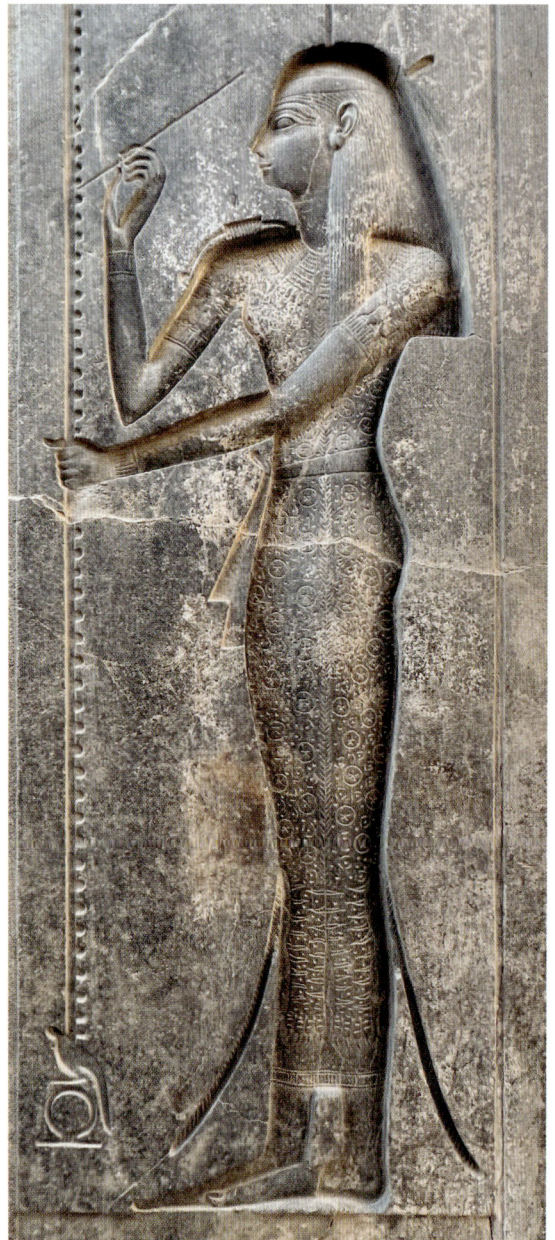
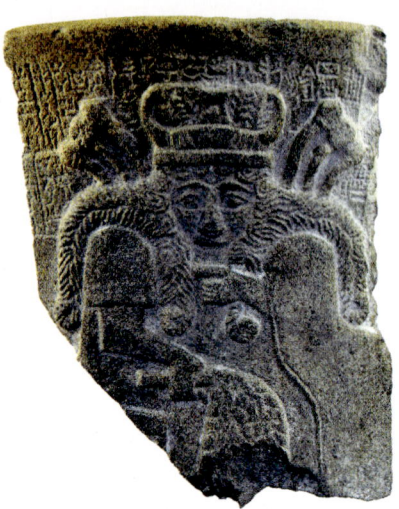

GODDESSES OF KNOWLEDGE | *Left:* A relief of the Egyptian goddess Seshat at the Luxor Temple, constructed around 1400 BCE. *Right:* A representation of the Sumerian goddess Nisaba, on the fragment of a 4,400-year-old vase from Iraq.

The high value that early societies placed on knowledge was not unique to the Greeks. This can be seen in the remarkable fact that gods of knowledge and wisdom are found in cultures around the globe. Many of these gods are actually *goddesses* whose dominion may also include writing, speech, crafts, or music.

In ancient Egyptian religion, Seshat is the goddess of knowledge, wisdom, and writing. She is closely associated with Thoth, another revered god of wisdom and writing. The Sumerian goddess of learning and writing is called Nisaba. For the ancient Greeks, wisdom, warfare, and crafts belonged to Athena. Saraswati is the Hindu goddess of knowledge, as well as music and the arts. The Yoruba people of West Africa call their goddess, or *orisha*, of wisdom Orunmila. In the Judeo-Christian tradition, wisdom is also personified as a woman; King Solomon says of her in the book of Proverbs, "She is more precious than rubies: and all the things thou canst desire are not to be compared unto her."

Returning to Western culture: the word "science" originates from the Latin *scientia*, meaning "knowledge." During much of the Middle Ages in Europe, this term generally stood for this broad idea. Specific disciplines such as anatomy, biology, chemistry, and physics were typically classified as *natural philosophy*, whereas what we now consider the "science" of astronomy was often grouped with the *mathematic arts* of geometry and music. The earliest universities in Europe, Africa, and elsewhere were established by religious institutions. The clergy were among the most well-educated, literate members of society. So, not only was it common for art and science to be entwined, religion was entangled with them as well.

During the Enlightenment, these disciplines came to be categorized in different ways. Fields of study once allowed to freely mingle—the sciences, arts, humanities, philosophy, religion—increasingly came to be separated. Astronomy, chemistry, physics, and mathematics were gathered under the umbrella of *science*, while painting, poetry, music, and so on were ushered into the *arts* camp. The new concept of *fine art* set apart certain categories of this creative work—sculpture and painting, for example—as greater in value and beauty than objects created through "craft."

Around the same time, European thinking about *science* also changed dramatically.

During the Enlightenment, a single approach became celebrated as the best means for studying and drawing conclusions about nature: intellectual reason. The belief emerged that through logic and objectivity, humans could attain the universal truth about the whole of reality. A popular idea of the time likened the cosmos to an enormous clock built by God and governed by the laws of physics. In this clockwork universe, everything could be broken down into its component parts, and these parts could be explained by science. This process would ultimately make it possible to predict how every aspect of the giant machine works. The conclusion: there is no question that science will not be able to answer.

From our perspective today, we can see how reductive, deterministic, and even presumptuous this assumption was. Not a few hundred years later, scientists themselves would be telling us what many artists and people of faith have known for a long time: It is impossible to know "reality" at its most fundamental level. Yet Enlightenment thinking still has a hold on our culture. The idea that science will eventually know all, and that objective truth is the best and perhaps the only real type of truth, hasn't gone away. The myths and gods of earlier, non-European cultures are often seen, at least among the strictly secular, as mere curiosities. But how about our present-day gods of knowledge?

If we do discover a theory of everything . . . it would be the ultimate triumph of human reason—for then we would truly know the mind of God.
— Stephen Hawking —

As a college freshman majoring in physics, I was excited that science might offer the prospect of catching even brief glimpses of the "mind of God." I was captivated by the idea of a clockwork universe, even as I troubled over what that would mean for free will. If it were possible to know the origin of the cosmos and predict its future—to discover a theory that would explain all of reality—it seemed to me that answers about humanity's role in the universe and even the meaning of life would naturally follow. At the same time, I recognized that wrapped up in this view of science were some unpleasant implicit assumptions. Such as the idea that nature is a sort of machine, that humans are separate from and above nature, that we have the right to conquer and control nature through technology.

My fledgling ideas about science were still muddled by pre-20th-century European thought. I hadn't yet taken a class on quantum physics.

Quantum physics is the science of how nature behaves at a very small scale, on the level of atoms, subatomic particles, and light. One of its fundamental teachings is that we—human beings—cannot separate ourselves from nature. We are in constant interaction with the world around us, and even the simple act of looking at a thing changes it and bonds us to it, so that we ourselves are changed. Everything must be considered in terms of relationships. Perfect objectivity is impossible.

Werner Heisenberg, one of the founders of modern quantum physics, cautioned against objectifying nature and even the concept of "objective reality." We cannot say anything about the behavior of nature separate from our manner of observing it. Heisenberg authored the famous *uncertainty principle*, which says that we cannot determine both the position and speed of a fundamental particle of nature with complete accuracy. For instance, suppose we want to know where an electron is located in a particular atom and how fast it's moving. The closer we get to pinpointing its speed, the fuzzier we are on its location. The best we can do is say where that electron *probably* is. Even if we could reduce human bias and use more precise instruments, we would not be able to eliminate this uncertainty.

Although quantum physics is the science of the microscopic, its principles hold true for the macroscopic world. Because of the effects of the interaction between the observer and the observed, we cannot fully pin down the nature of "reality." Once we fully appreciate this, it becomes clear that many of the dichotomies that have made a home in our thinking are mere houses of cards. As Heisenberg wrote, "The familiar classification of the world into subject and object, inner and outer world, body and soul, somehow no longer quite applies, and indeed leads to difficulties." In the same essay, he states that the mathematical formulas and theories of science do not describe nature per se, but rather our *knowledge* of nature.

Modern quantum physics was born over a century ago, but its ideas have not penetrated deeply into Western culture. The implications of this science— that we are intimately connected to the universe and one another—have not been internalized. Many people, some scientists included, still believe that the goal of science is to obtain truth about "objective reality" by means of a strictly logical approach. This belief forms part of a larger "myth of objectivism" described by George Lakoff and Mark Johnson in their classic work *Metaphors We Live By*. Here's the essence of the myth: "Science can ultimately give a correct, definitive, and general account of reality, and, through its methodology, it is constantly progressing toward that goal." All we have to do is eliminate every possible source of irrationality, emotion, bias, and other subjective influence that can interfere with science. Reason is all.

The flip side of the myth of objectivity is the myth of *subjectivity*. According to this story, our most reliable beacons in life are our feelings, intuition, imagination, and spiritual awareness, which are entirely subjective. These qualities transcend objectivity, a cold, abstract, indifferent approach to life that in the end cannot help us meet our most essential needs and desires. In this view, the arts, with their capacity for personal expression and inspiration, penetrate to the reality of things far better than science does.

But subjectivity, too, can be taken to an extreme—everything becomes a matter of individual taste. Anything can be labeled as art—a blank canvas, a painting created by a rabbit, a sonnet composed by artificial intelligence—as long as it is provocative, spurs a debate, or someone says it is so. The standards for what constitutes good or important art evaporate. Like the Cole Porter song, *anything goes*!

It's not only science that is made poorer by the myths of objectivity and subjectivity. In our world of hard oppositions, art gets the worst of it. In a dichotomy, such as science versus art, usually one concept dominates. "Separate but equal" is a rarity indeed. Man dominates nature, logic overrides emotion, body conquers soul, white beats black. Scientific knowledge trumps spiritual knowledge. Science is valued more highly than art.

Two brief examples indicate how, in the United States, art is undervalued compared to science. In 2020, the US government allocated $162.5 million to the National Endowment for the Arts, just a tiny fraction of the federal budget. Meanwhile, $717 *billion* went to research and development in science—more than *4,000 times* the spending on art! On a local level, when public schools make tough decisions about allocating diminished resources, which programs are typically the first to go? The art and music classes.

A first step to restoring balance is to recognize the falseness of the dichotomies we've inherited. We may not have created them, but we have a choice as to whether to perpetuate them. There is always a third way.

The greatest scientists are artists as well.
— Albert Einstein —

What's needed is a synthesis of the objectivity and subjectivity. As individuals, we all rely on both reason and intuition, both logic and feeling to some degree. The same for the scientist and the artist. Making art requires certain special skills and technique. The painter must know how colors work. The musician must know how to put together notes, harmonies, and rhythm. *There is a science to art.*

The scientist, on the other hand, cultivates the ability to recognize patterns in nature and then uses her imagination to come up with good explanations for them. The physicist learns to trust her gut when something about a theory doesn't quite add up. She learns how to ask fruitful questions and follow a hunch. *There is an art to science.*

So, what *is* art? What *is* science? Given that art and science have meant different things to different people at different times in history, is it even possible to define them in a coherent way? I think so. As long as we remember that our definitions are particular to our specific place and moment, defining art and science can be beneficial.

Here are a few ways of thinking about art and science that I find enlightening. The first way regards art and science as *systems*—collections of things organized in a specific way, by humans and for humans. The dominant system of art in Western society consists of certain categories of disciplines (such as visual art, poetry, music, dance), institutions (such as schools, museums, concert halls), people in certain roles (such as artists, critics, patrons), traditions (such as the emphasis on European art history, commodification), and values (such as creativity, imagination, novelty, individualism). The system of science has some similar categories: its own network of institutions (universities, research organizations, industry), roles (scientists, engineers, funders), traditions (the emphasis on European contributions to science), and values (such as objectivity, innovation, individualism).

When we call people artists or scientists, it's typically because they are connected to one of these systems. This way of thinking about art and science is practical, but it has limitations. A system imposes rules, spoken or unspoken, on its participants. It can keep art and science locked in their silos and leave out a lot of people. Just as one can be an educated person outside the educational system—say, like Frederick Douglass—one can be an artist or scientist outside the dominant systems—as the astronomer Benjamin Banneker and the painter Clementine Hunter proved.

Another way of looking at science and art has to do with *ways of thinking and of approaching life*. As Nina Simone said, "Jazz is not just music, it is a way of life, it is a way of being, a way of thinking." I think the same can be said for art in general, and for science, too. Painting, for me, is not just art; it's a way of seeing the world. The habit of painting has focused my attention on color, light, nature, personalities, stories. It has become a language that I think in, and it affects how I express myself—and not just when I'm applying paint to canvas. Similarly, astronomy is not just science for me. It has become part of my understanding of the world; it is integral to my belief that we are all connected and that the universe is full of infinite possibility.

I like to think of the space surrounding art and science as a spectrum. Not a linear spectrum, like the colors of the rainbow, but one that curves, twists, and wraps back upon itself—like the merging colors of a soap bubble. At one place in the spectrum, art is weighted toward emotion and feeling. At another spot, science is weighted toward logical understanding. The scientist approaches the universe in a manner that stresses logic, what can be measured, and what can be quantified in the material world. She looks for physical evidence to support her theories. The artist approaches nature in a way that emphasizes personal meaning and expression. She is sensitive to spiritual proof.

The spectrum metaphor allows us to see similarities in the ways artists and scientists approach things. Both value beauty. Their work requires creativity, curiosity, critical thinking, technical skill, and imagination, with the joy of discovery being one of their common rewards. This metaphor also shows that art and science are interdependent. Art doesn't exist without science. The tools needed to make art—paint, paper, musical instruments, photographic film—result from the application of scientific knowledge.

In a sense, science doesn't exist without art, though it might be harder to see how. Art's influence often comes in the elusive form of *inspiration*, which is indirect and impossible to quantify. Art can help us to see the world in new ways, shift our perspective, and literally change our thinking, which may help us solve difficult problems. Neuroscientists have confirmed the transformative power of art, showing that observing *and making* art can create new neural pathways in the brain and stimulate imagination. Would Einstein have come up with the general theory of relativity in the manner he did had he not been playing Mozart on the violin while he was working it out? Would Satyendra Nath Bose (for whom the subatomic particle *boson* is named) have made the discoveries that laid the foundation of statistical quantum physics had he not also been a classically trained musician who played and composed on the esraj? More to the point, would Einstein and Bose have even been motivated to do their work as scientists if they lived in a world without art? Science is necessary for life, but art makes it worth living.

A third way of defining art and science is based on *ideal roles*. I say "ideal" because that imposes a high standard. We define a "pillow" based on our mind's ideal picture of what a pillow should be. You can lay your noggin on a cinder block if you're out of other options, but it would not meet your expectations of ideal pillowhood. We might benefit from developing some new ideal standards for art and science. I'd like to offer some suggestions.

Science is a methodical process of asking meaningful questions that lead to a better understanding of the physical world. By this definition, medicine women and midwives are scientists. Physics is a science concerned with how nature works at the fundamental level. It seeks to discover the laws governing the behavior of the universe. To tease out these laws, it deals with phenomena that can be measured and quantified, with theories that can be tested experimentally. The focus of this book is astronomy, the science that uses light and images to explore the workings of the cosmos. *Astrophysics* has the goal of understanding the physics of the entire universe itself. Today the terms "astronomy" and "astrophysics" are used interchangeably.

Yet science is more than a systematic pursuit of knowledge about the physical world. Ideally, it is a pursuit of *wisdom* that teaches us how to live in harmony with the world and be good stewards of it. The highest science liberates and serves humanity and contributes to a beautiful quality of life.

Art, the human creative activity intended to communicate *feeling*, can communicate other information and ideas, too. But these are secondary. The French post-impressionist painter Paul Cézanne had it right when he said, "An art which isn't based on feeling isn't an art at all . . . Feeling is the principle, the beginning and the end; craft, objective, technique—all these are in the middle."

When emotion is successfully communicated, something magical happens: connection. The Russian writer Leo Tolstoy, a contemporary of Cézanne, said in *What Is Art?*, "Art, all art, has this characteristic, that it unites people." Art is about human connection, and the main glue it uses to bond us is feeling.

Art encompasses far more than we typically think it does—and far less than what is delivered for our consumption in popular media. A church lady

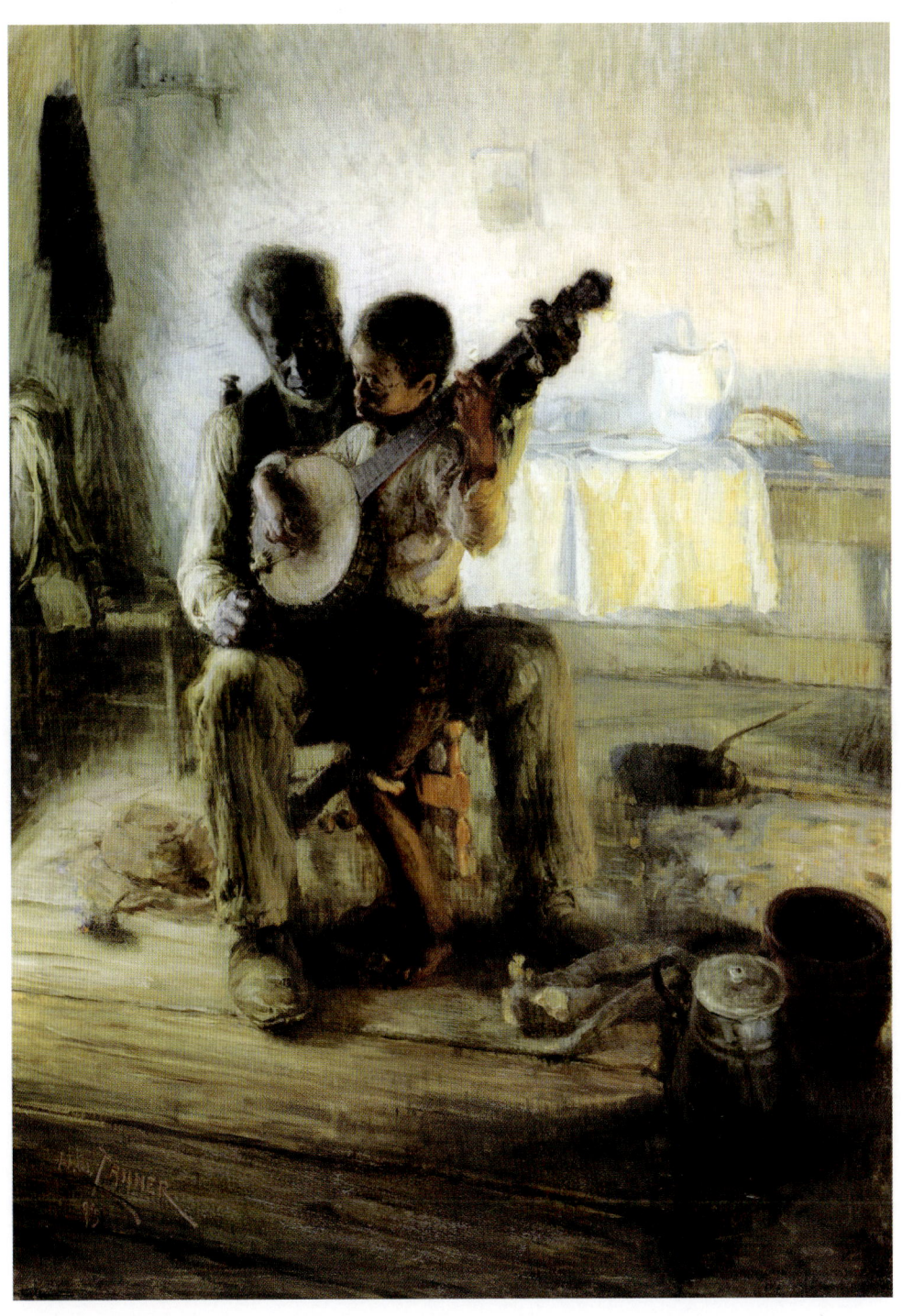

THE BANJO LESSON | Henry Ossawa Tanner, 1893, oil on canvas, 49 × 35½ in (124.5 × 90.2 cm), Hampton University Museum, Hampton, Virginia.

dressed in an original outfit of her own design, a grandfather improvising a song with his grandchild, a child telling a story—all are creators of living works of art.

Surely you've had the experience of reading a book or listening to a song, and you felt as though the artist were speaking directly to you. They said something you knew to be true in a way that really hit home. Before then, you hadn't been able to articulate it yourself. The experience opened a new window for you; your world expanded and you felt more connected. This is the power of art. Not only did you become connected to the artist, but you instantly formed a relationship with others who read that book or listened to that song or shared a similar human experience. Art enhances our humanity. The highest art unifies.

Since new developments are the products of a creative mind, we must therefore stimulate and encourage that type of mind in every way possible.
— George Washington Carver —

George Washington Carver once said, "Reading about nature is fine, but if a person walks in the woods and listens carefully, he can learn more than what is in books, for they speak with the voice of God."

Carver was a living bridge between different definitions of art and science. Born just before the abolition of slavery, through a combination of providence and an unrelenting determination to pursue knowledge, Carver was able to attend Iowa State University, where he studied painting, music, botany, and agriculture. After graduation, he became the college's first black professor and was soon invited by educator and author Booker T. Washington to come to what is now Tuskegee University; there, he would teach and conduct research for nearly five decades. Carver was certainly an active member of the *systems* of art and science as they took form in early 20th-century America. As a formerly enslaved person living through the Jim Crow era, he also existed outside the system. The unique circumstances of his life, including exposure

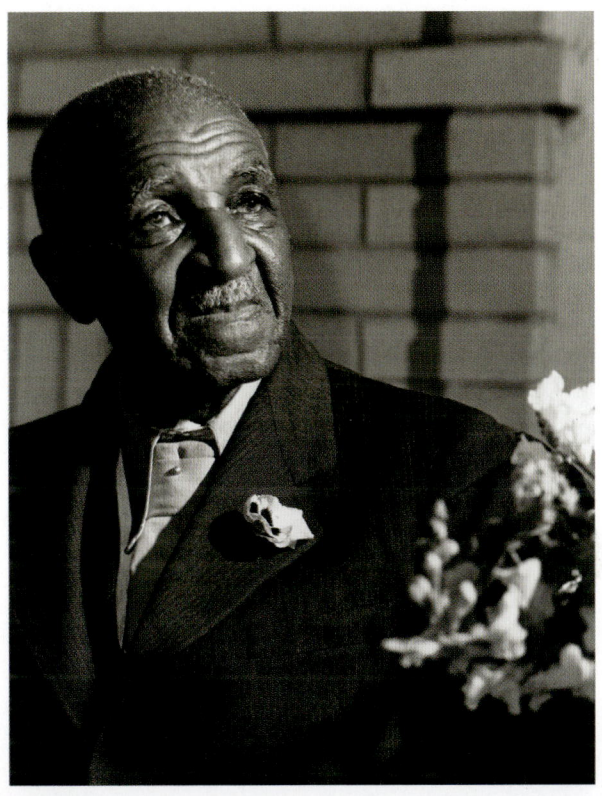

GEORGE WASHINGTON CARVER | The great scientist and artist in 1941 in Tuskegee, Alabama.

to nature as a child and his passionate faith in God, led to the development of a personality that effortlessly blended art, science, and spirituality. Though he was persuaded to pursue botany in college, Carver's first love was art, and he painted throughout his life. (He even created paints from peanuts and used them in his own artwork!) He recognized no divide between science and art, saw no contradiction between science and faith—he tapped into any source he could that would allow him to hear "the voice of God."

Above all, Carver was interested in using his creative power to serve humanity. The impetus behind his prolific innovations, which ultimately transformed American agriculture, was to raise the standard of living of poor black farmers. Hence his achievements in crop rotation and biochemical engineering. Understanding that many of the people he wanted to reach were illiterate, he used his talent as an artist to illustrate his ideas, showing rural Southern black people better ways to cultivate the land and grow foods that would improve their diets.

By all accounts, his was a gentle and inquisitive spirit. He was fond of quoting a poem by Alfred, Lord Tennyson. Its lines are a touching metaphor for how the mysteries of life begin to reveal themselves when we seek holistic understanding.

> Flower in the crannied wall,
> I pluck you out of the crannies,
> I hold you here, root and all, in my hand,
> Little flower—but if I could understand
> What you are, root and all, all in all,
> I should know what God and man is.

A reminder from quantum physics: The divisions characterizing much of our world are based on false ideas. But there's an even deeper implication: *We create our own reality.* This is something that many artists, thinkers, spiritual leaders, and wise grandmothers have known for ages. Quantum physics reveals that the mere act of *observing* something forces it to adopt a certain state. Before we observed it, that thing existed in the mysterious territory of infinite possibility. In other words, we can't help but change the world. It's up to us whether we influence our environment as passive occupants or active creators working toward the world we would like to see. That's our choice.

In the act of trying to understand the universe—whether through science or art, theology or philosophy—we leave our mark on it. In the pages that follow, we will discover artists and scientists who, in leaving their mark, enlarged humanity's vision. We will explore different approaches to viewing the universe that, when united, magnify our ability to see, create, and connect. When we try to comprehend this spectacular entity, we are at once artists and scientists *painting the cosmos.*

CHAPTER TWO

SKY FULL OF STORIES

Each day has a story to—deserves to be told, because we are made of stories. I mean, scientists say that human beings are made of atoms, but a little bird told me that we are also made of stories.
— *Eduardo Galeano* —

Every story I create, creates me. I write to create myself.
— *Octavia Butler* —

The internal machinery of life, the chemistry of the parts, is something beautiful. And it turns out that all life is interconnected with all other life.
— *Richard Feynman* —

I f you've had the opportunity to look up at a dark, star-filled sky, or even if you've only seen photos of the Milky Way, you're familiar with the sense of wonder and mystery that the cosmos never fails to evoke. The sky is boundless, intangible. What is that array of lights up above? Why do they behave the way they do? How did they get there? What is humanity's relationship to it all?

Imagine stepping back in time 2,000 years, or just 200. How would you go about searching for answers to these questions? There's a good chance you don't have easy access to books. TV doesn't exist. And (horrors!) no cell phones. Not that having one would help, since the internet hasn't been invented yet. What would you make of those brilliant lights strewn so generously across the night sky? How about that hot, big light that dominates the day and constantly changes size and color? What story would you tell to make sense of it all?

Today, because of light pollution, especially in the Northern Hemisphere, where 90 percent of the world's population lives, the sight of a star-spattered sky is much less common than it was for most of our ancestors. A typical suburbanite may be able to see a few hundred stars a night; someone living in a big city like New York would be lucky to make out the handful of stars that manage to penetrate the glare of artificial lights. But for most of human history, a star-filled night sky was a constant, palpable presence in people's lives.

Our ancestors used the Sun, Moon, and stars to mark the passage of time and develop calendars. Keeping a vigilant watch over the passage of the seasons was essential for agriculture, rituals, holy festivals, navigation, and commerce. Painstaking observations of the sky, made year after year and century after century, revealed patterns. Based on these patterns, our ancestors make complex calculations to predict future events. What's more, the sky was a source of inspiration. Out of the fundamental human urge to create meaning and connection, they invented stories about the origin of the cosmos and humanity's place in it.

Around the globe, a common theme of creation stories is how light first entered a universe that previously slumbered in darkness. From ancient Egyptian stories about Ma'at—the daughter of Ra, the sun god, who along with her father rose out of the primordial darkness and chaos to bring truth, harmony, and order into the world—to the biblical account of God speaking light into existence, humans have conceived stories that reveal our place in the universe. Stories help us make sense out of chaos, to find order and safety in an uncertain existence. We tell and retell them, not only as a way to establish our relationship with the cosmos but to preserve our connections to past and future generations.

Long ago, the world was in darkness, and the night sky was pitch-black because there were no stars. One night, a lonely little girl decided she wanted to visit other people. So she took some embers from the fire and threw them upward into the sky. They stayed there forever as the glowing band of light that we in the West know as the Milky Way.

The Khoisan are among the first people to inhabit lands that are now parts of Botswana and South Africa. They have lived there for thousands of years, as nomadic cattle herders and hunter-gatherers, and archaeologists have found artwork there that they believe to be among the oldest in the world. Over the generations, the Khoisan have passed down this story, which contains many truths that we all can recognize. It tells us that although the stars seem to be permanent, like everything else, they had a beginning, and their existence is connected to our own. This simple tale also grapples with the desire not to be alone. The little girl is inspired to light up the sky so that she can find other people—similar to how the ancient Egyptian god Atum, perceiving his solitude, fathered Shu and Tefnut, the gods from whom all else was created.

According to the creation story of modern Western science, once, billions of years ago, the universe was in fact dark. Light wasn't free to shine until 380,000 years after the Big Bang. It took another 200 million years or so for the first stars to form and light up the sky.

The stories we tell about ourselves give meaning to our lives. Storytelling, perhaps humanity's oldest art form, is a creative, emotional response to our most fundamental questions: *who, why, what,* and *how?* How did we

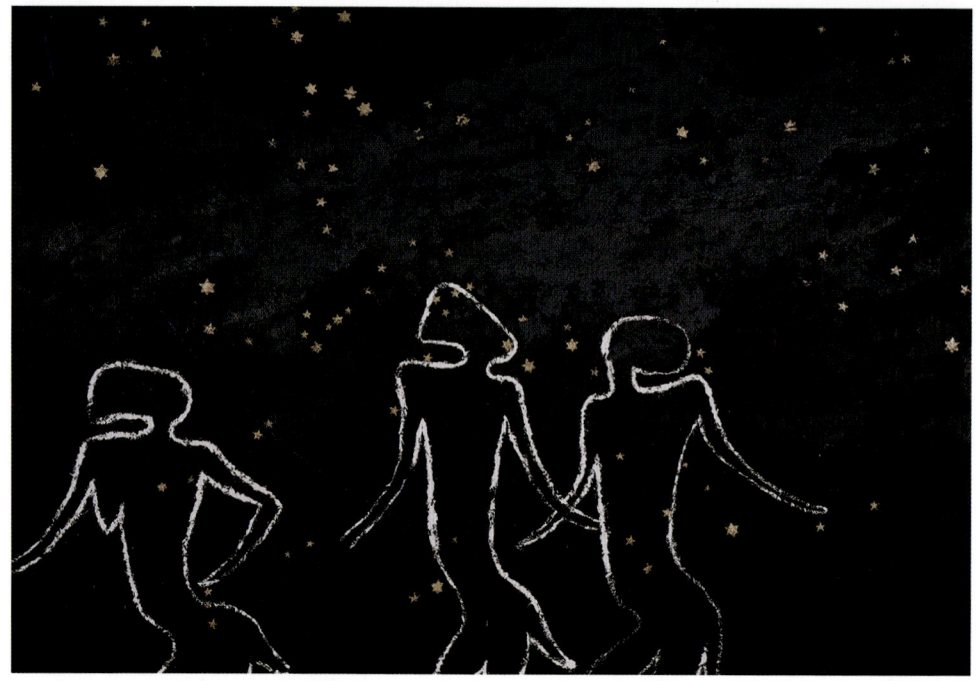

UNTITLED | Gavin Jantjes, 1989—90, acrylic on canvas, 78¾ × 118⅛ in (200 × 300 cm). From the series titled *Zulu*, which means "sky" or "heavens" in the Zulu language, this painting is inspired by the Khoisan story about the creation of the Milky Way.

get here? What is our destination? How does *my* individual story fit into the whole? Individually and collectively—as families, communities, cultures, and nations—the stories we tell have the ability to illuminate our reality and give us new understanding. But more than that, stories have the power to create *new* reality and to unite us, as we confront the hard, frequently beautiful, and endlessly surprising reality of the human condition.

Light, the Universe's Silent Storyteller

A story may be told using words, paint, wood, dance, or music. The universe uses light as one of the ways it tells its stories, which are intimately connected with humanity's own. Light travels from the near and far reaches of the cosmos, carrying a tale of where it's been and the journey it's been on along the way. Light connects us with the cosmos, bearing the message that everything in it is interconnected.

Light tells us that the universe is dynamic and ever changing. Yet patterns exist. An obvious one is what appears to us as the Sun's east-to-west journey across the sky every day due to Earth's spin around its axis. Similarly, the stars rise in the east and set in the west each night. As Earth orbits the Sun, the appearance of the night sky gradually changes as constellations disappear behind the Sun and ultimately reappear.

Our lack of depth perception when looking up at the vastness of interstellar space makes it seem as though the stars all lie at the same distance from us and are securely pinned to the firmament. Each night they appear to rotate across the sky in a fixed group. Yet there are a handful of bright lights that don't stay fixed with respect to the stars or to each other. Our ancestors noticed this. They saw that night after night, these special lights meander across the sky and sometimes, for weeks or months at a time, disappear altogether. They appear to *wander* among the stars. Thus, the word for "planet" in English and several other European languages derives from the Greek *planētai*, meaning "wanderer."

The five planets closest to Earth—Mercury, Venus, Mars, Jupiter, and Saturn—have been known since antiquity. The outermost planets, Uranus and Neptune, were not discovered until 1781 and 1846, long after the invention of the telescope.

All eight planets were born 4.6 billion years ago, from the same primordial cloud of gas and stardust that gave us the Sun. Our closest star, the Sun is by far the biggest, most massive body in the Solar System.* Because of its overwhelming gravity, the planets are anchored to it, ceaselessly falling into orbit around it in nearly the same plane, as they receive and reflect its light. Like all other stars, the Sun derives its power from nuclear fusion, the process by which atoms join together and release light. Because planets and the Moon are much less massive than stars and, therefore, have much weaker gravity, they don't shine the way that stars do. Stars are self-luminous. Planets, by contrast, reflect the light of the Sun.

Venus was a significant planet for many ancient societies. The third naturally brightest celestial body, after the Sun and Moon, it is bright enough to see in the daytime and to cast shadows at night. Before the telescope was invented, the Maya of Mesoamerica developed a sophisticated astronomy that included precise calculations of the planet's movements, reaching far into the past and forward into the future. The Maya correctly measured its synodic period—the time it takes for Venus to return to the same place in its orbit as viewed from Earth—to be 584 days. Venus was central to Mayan culture, so much so that the Maya identified it with Kukulkan, god of the sky. Its movements informed their architecture, calendars, and the timing of spiritually significant ceremonies.

The famous Dresden Codex, a Mayan astronomy book from the 11th or 12th century, is believed to be a copy of a manuscript at least 200 years older. It contains meticulous astronomical charts of Venus and other celestial bodies, including the Moon and Mars. One of the oldest surviving texts of the Americas, it was stolen during the Spanish conquest of the 1500s and wound

---

* A reminder from high school physics: Mass is the amount of matter in an object. This is not the same as weight, which is the measure of the force of gravity on an object.

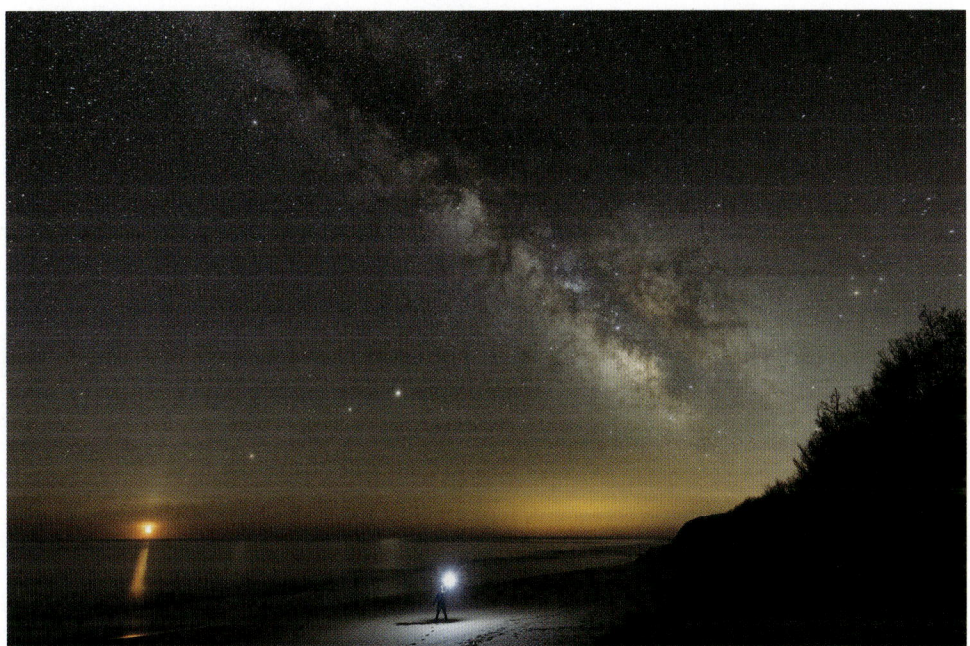

THE MOON, MARS, SATURN, AND JUPITER TOGETHER | Because the planets all revolve around the Sun in nearly the same plane, they can sometimes line up in the sky. This photo was taken at dawn from the Bulgarian Black Sea Coast. Against the background of Milky Way stars, it shows the Moon (bottom left), as well as Mars, Saturn, and Jupiter extending upward toward the right.

up in Germany in the 1700s, where it was eventually bought from a private collector by the Royal Library of Dresden.

We Are Connected to the Past

Light tells the story of the Sun, the colossus at the center of the Solar System that keeps Earth locked in its orbit at a distance of roughly 93 million miles (150 million kilometers). To get a sense of how far away this is, imagine squeezing the Sun into the size of a grapefruit, with Earth and everything in the Solar System scaled down proportionally. Earth would lie at a distance of about 16.4 yards (15 meters) from this grapefruit-sized Sun, and it would be roughly the size of the period at the end of this sentence. Here's another way

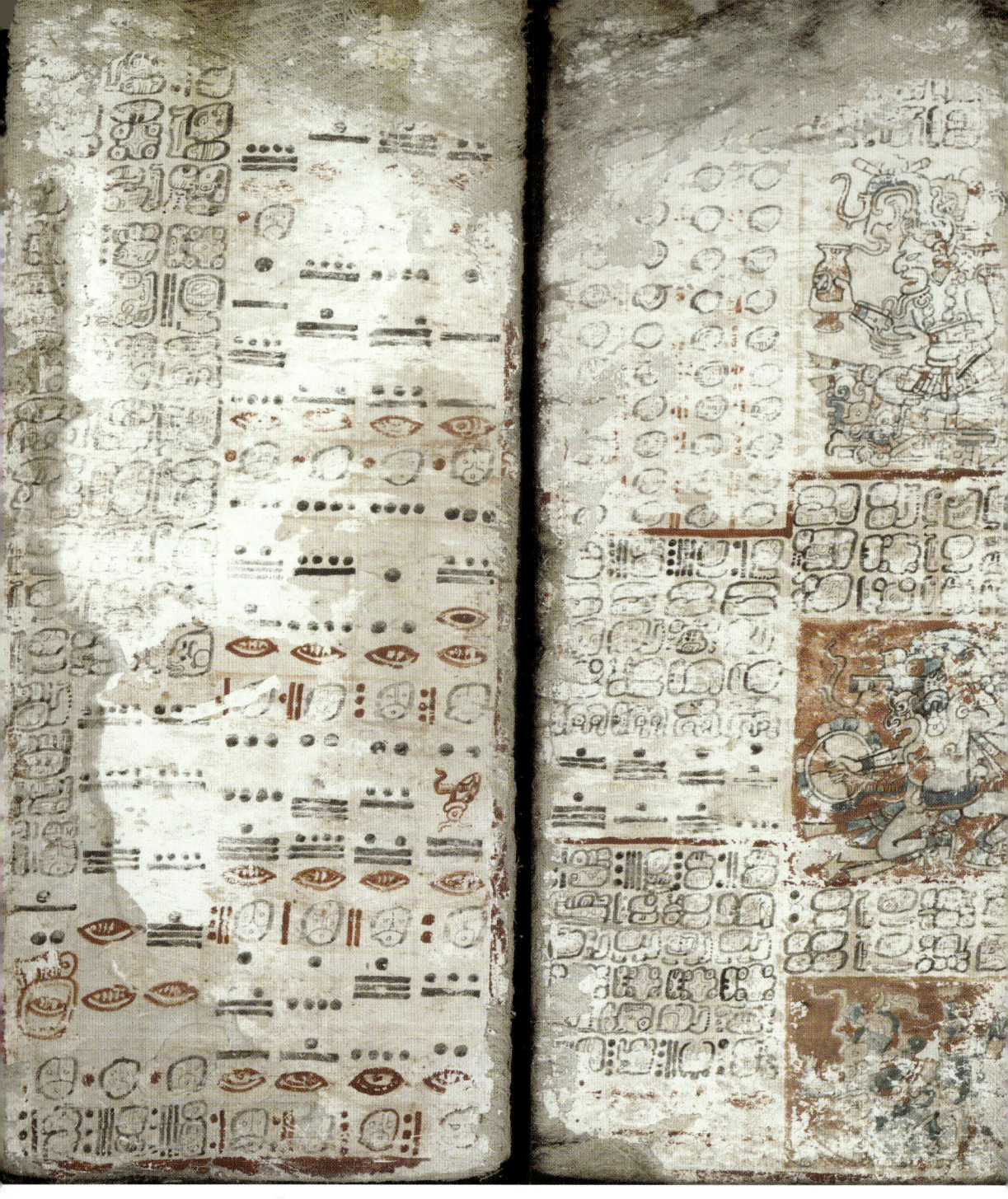

THE DRESDEN CODEX | Preface of the Venus Table in the Dresden Codex (first panel on left), and the first three pages of the table. The table tracks the movements of Venus as seen from Earth. The three paintings on each page may depict deities associated with Venus. Numbers are denoted using the Maya dot-bar numerical notation. Other symbols represent information such as the names of days and the cardinal directions.

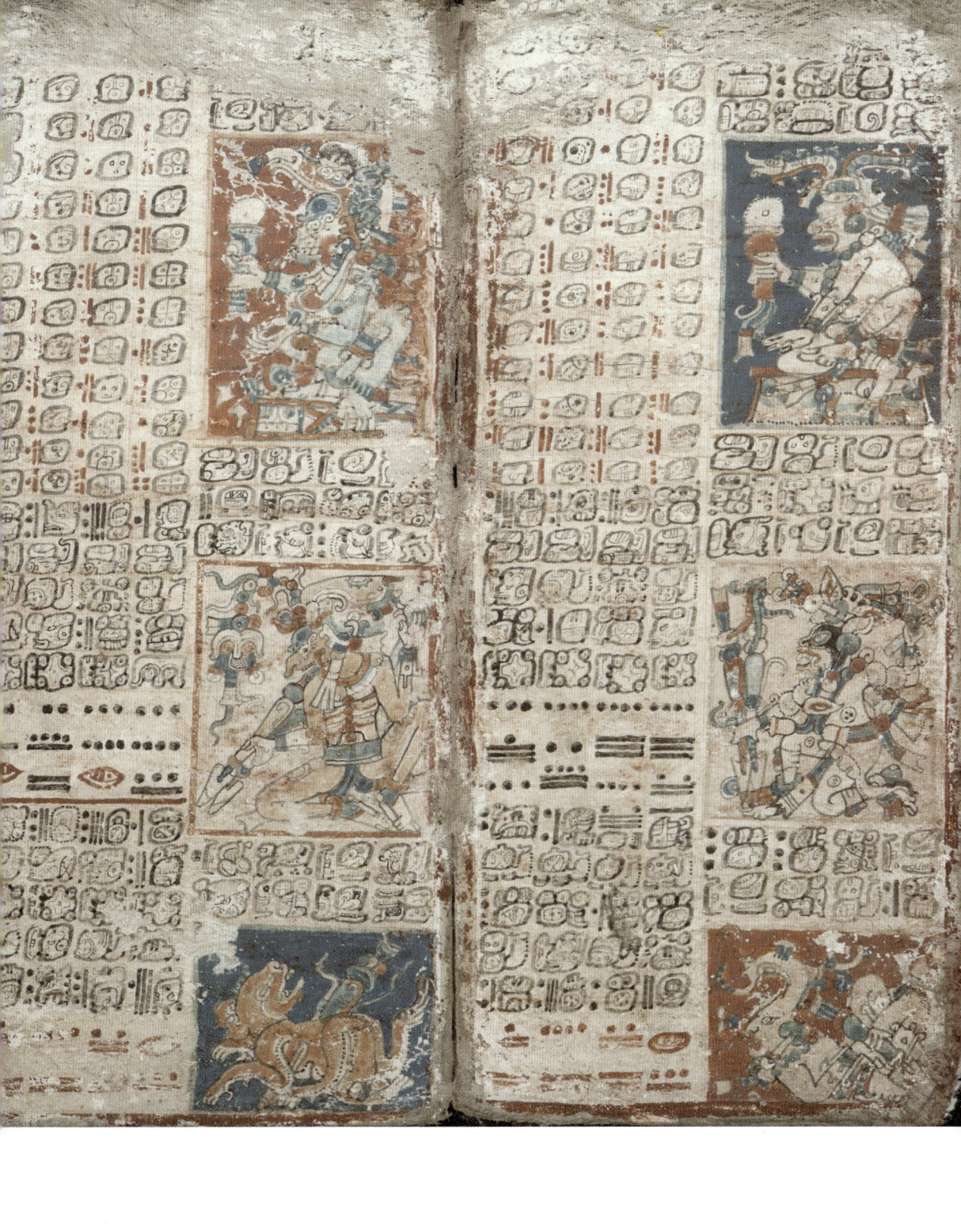

to consider the distance. If we were to travel toward the Sun in a spaceship going 20,000 miles per hour, it would take just over six months to reach it. But when light leaves the surface of the Sun, it takes a mere eight minutes and 16 seconds to traverse interplanetary space and arrive on Earth. Since Venus is much closer to Earth, it takes only three minutes and 46 seconds for the light it reflects to reach us. From the Moon, which is even closer, light makes the trip in just 1.3 seconds.

Now let's step beyond the Solar System into interstellar space. The first stars we encounter are a family called Alpha Centauri. The two biggest stars in the family are similar to the Sun in mass and luminosity. The third, Proxima Centauri, has only 12 percent of the mass of the Sun and shines nearly a thousand times dimmer. The three stars are locked in close orbit by their mutual gravity, and they are so distant from us that the two Sun-like stars, Alpha Centauri A and B, appear as a single star. Proxima is too faint to be seen with the unaided eye.

Here's how Alpha Centauri fits into our Sun-as-grapefruit analogy. Its distance from the Solar System would be nearly 1,860 miles (3,000 kilometers). In the actual universe, that distance is about 25 trillion miles. This is an insanely large number! And quite difficult to grasp. Because the distances in space are so vast, it gets cumbersome to speak in terms of our standard units of measures, such as miles or kilometers. The *light-year*, which is the *distance* that light travels in one year, is more convenient. Light travels at a speed of 186,000 miles per second. All the time, everywhere in the cosmos. It's the fastest thing in the universe and covers enormous distances quickly. So when we talk about the distance to Alpha Centauri, rather than saying it's 25 trillion miles away, we can say it's 4.25 light-years away.

This means that if you were to go out tonight to look for Alpha Centauri—visible from Earth's Southern Hemisphere—the light landing on your eyes began its journey from the surface of those stars more than four years ago.

This illustrates a couple of important ideas. Despite its speed, it does take time for light to get from point A to point B. This means that the farther away something is, the longer it takes for its light to reach us, and more distant objects appear younger than nearby objects. We see Alpha Centauri as it was four years ago. The Sun is closer; we see it as it was only 8.3 minutes ago. (This is equally true here on Earth, though the timescales are far too short for our brains to register them. If you would like to appear younger to your crush, you'd have to stand really, *really* far away!) *We always see things as they existed in the past. There is no "now."* Because it takes time for light to deliver its message, more distant objects provide us with glimpses into the past. Astronomers use this remarkable property of nature to construct a history—a story—of the cosmos.

Humanity's story is intimately connected with the light of the stars. Sirius, the brightest star in the night sky, has been significant to cultures around the world. A bluish-white star about twice as far away as Alpha Centauri, Sirius is nevertheless brighter; it's more than 25 times more luminous than the Sun. You can find it in the night sky by following the three bright stars in Orion's belt southward.

In ancient Egyptian society, one of the most important yearly events was the flooding of the Nile River, which happens at roughly the same time of year as the heliacal rising of Sirius, or the first day in July or August when this star would appear on the eastern horizon just before sunrise. By forecasting this day, farmers could plan their sowing and harvesting seasons around the predictable flooding cycle. And they could take advantage of the natural fertilization that would occur along the riverbank when the water rose.

Sirius happens to be one of the closest and most luminous of several thousand stars that can be seen with the naked eye, without a telescope. These stars all lie within a few hundred light-years of Earth—our celestial backyard, by cosmic standards. They account for only a drop in the vast galactic pond that is the Milky Way, which contains more than 200 *billion* stars. Separated by immense stretches of space, most of these stars occupy a region shaped like a vast disk, somewhat like a vinyl record. Wrapping around the disk are spiral arms that contain the youngest, brightest stars in the Galaxy. We think the Milky Way has two major spiral arms, with a number of smaller ones radiating from them. Our Solar System is located in an inconspicuous suburb roughly 30,000 light-years from the center of the Galaxy, slightly above the middle of the disk in a minor arm called the Orion Arm.

How the night sky looks is determined by our vantage point from within the Galaxy. When the ancient Romans looked up at the broad, fragmented stroke of light across the night sky, they saw a road made of milk—*Via Lactea*, Latin for "Milky Way." And the word "galaxy" comes from the Greek word *galaxias*, meaning "milky."

The different names that cultures have given to humanity's galactic home tell something of their understanding of the cosmos. Many of these names allude to a path or river: Hay Merchants Way (Arabic), Silver River (Vietnamese, Chinese), River of Heaven (Japanese), and Way of Birds (Estonian). The aboriginal people of Australia see a great emu in the sky—the long, slender neck and body of the large ostrichlike bird are drawn by the dark lanes of stardust stretching across the plane of the Milky Way. Various groups in southern Africa refer to it as the Sky's Spine or God's Backbone, suggesting that the Milky Way is a structure that gives support to the sky.

All of these names describe how our Galaxy looks from our perspective within the disk. The broad, luminous streak across the sky that looks like a path, river, or backbone is actually the merged light of millions of stars in the disk, so far away and crowded so closely together on the sky they cannot be individually distinguished. If our ancestors could have had a bird's-eye view of the Galaxy and seen its grand design spiral arms, I imagine that today we might be calling it the Eagle's Talons, the Silver Spider, or perhaps God's Pinwheel.

The dark lanes carving through the Milky Way, or the Great Emu, do not indicate the absence of stars. They are caused by vast clouds of stardust. These fine particles of dust are typically hundreds of times smaller than the width of a human hair, and they are composed of atoms like carbon, oxygen, silicon, and iron. Stardust permeates space and acts as a screen, blocking the light from stars in the background. When the astronomer William Herschel was observing the sky in the late 1700s, he was alarmed to discover an apparently starless region. "Here is truly a hole in heaven!" he remarked. Today we know that these "holes" in the sky are actually clouds of *dust*, the material out of which planets are born.

Our ancestors not only named the patterns they observed up above but also created maps of the sky. A 300-year-old star chart created by the Skidi Pawnee people, originally from the area we now know as Nebraska, shows many familiar groupings of stars, including the Big and Little Dippers, Corona Borealis, the Hyades, and the Coma and Pleiades star clusters. They are represented in a highly abstract manner that is as artistic as it is scientific. Stained onto a tanned elk skin, the map—now housed at the Field Museum in Chicago—was more than a utilitarian handicraft; it reflected sacred knowledge, spiritual and cultural values that could be transmitted down the generations. The Skidi Pawnee, like many other peoples, developed calendars, navigation, ceremonies, and rituals based on the stars. They even organized their villages according to patterns they observed in the night sky. Their way of life reflected their belief that they came from the stars.

The Skidi Pawnee were not wrong. Modern Western astrophysics tells us that the building blocks for life as we know it, including human life, were forged billions of years ago inside the hearts of stars.

We Are Connected to the Stars

Right this moment, somewhere in the universe, a star is being born. Every second, in fact, nearly 5,000 stars begin illuminating pockets of the cosmos with their new light. That's 400 million stars every day—more than three times the number of babies born in the world every year. Just as all babies are born by women, all stars emerge from one source: interstellar clouds of gas and stardust called *stellar nurseries*.

And at this very moment, somewhere in the universe, a star is dying. Every second or two, somewhere in space, a star says its final goodbye in a spectacular display of cosmic fireworks. Stellar birth is much more common than stellar death. In the Milky Way, the equivalent of one to three solar masses of stars is born every year, while star death in the Galaxy occurs only two or three times a century.* Star birth is relatively inconspicuous, occurring in the silent obscurity of stellar nurseries. Star death, by contrast, is explosive, ostentatious, and unforgettable.

---

* A "solar mass" is the mass of our Sun, about 2×10^{30} kilograms (that is, 2 nonillion kilograms).

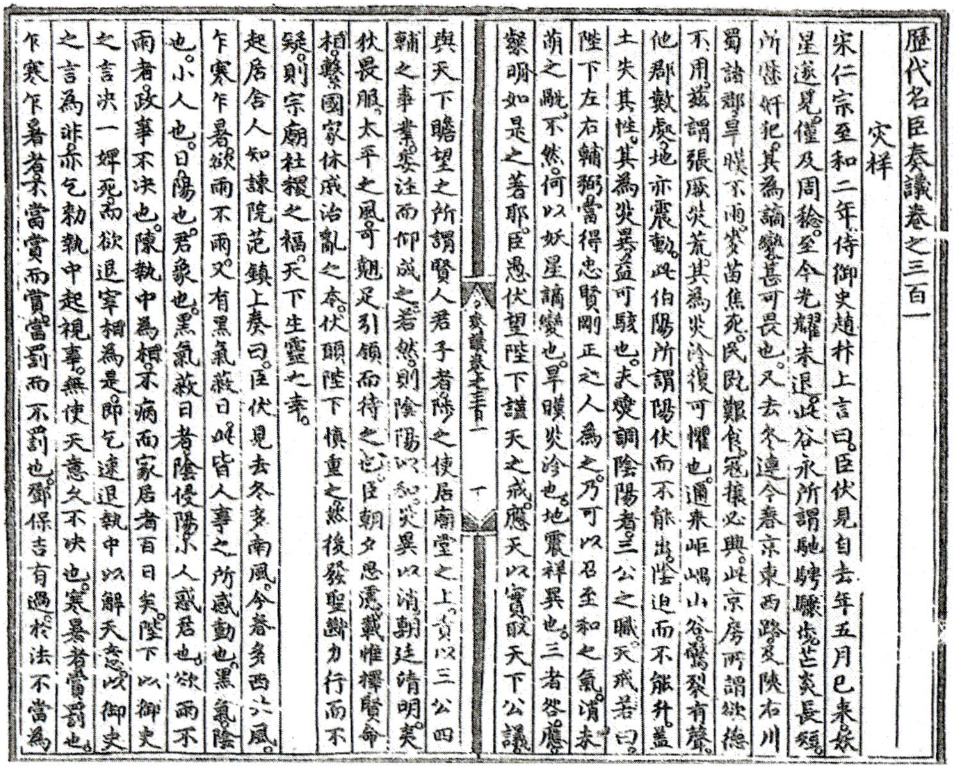

歷代名臣奏議卷之三百一

災祥

宋仁宗至和二年侍御史趙抃上言曰臣伏見自去年五月巳來妖星遂見僅及周稔至今光耀未退迨谷永所謂馳騁驟驟虎芒長頍所繫奸邪犯其為讒變甚可長也又去冬遼令春京東西路夏陝右川蜀諸鄰阜膜不雨炎茵焦死民飢艱食冦攘必興此京房所謂欲德不用謂張厭災荒其為災變益可駭也他郡數處地亦震動此伯陽所謂陰陽所謂陽伏而不能出陰迫而不能升地震祥異也夫燮調陰陽者三公之職天戒若曰土失其性其為災暴益可駭也乃可以召至和之氣應天以實敗天下公議

陛下左右輔弼當得忠賢剛正之人為之乃可以召至和之氣應天以實敗天下公議

萌之既不然何以妖星讒變也旱暵炎冷也地震異也三者咎應天以實敗天下公議

燮明如是之著耶臣愚伏望陛下謹天之戒應天以實敗天下公議

與天下瞻望之所謂賢人君子者陽之使居廟堂之上貞以三公四輔之事業注而仰成之若然則陰陽以知災異以消朝廷靖明矣伏望太平之風可翹足引領而待之忠臣朝夕思憑戴惟擇賢命也小人也日陽也陰象也君象也黑氣救日者陰侵陽小人感君也欲退寧揣為是師乞速退執中以解天忈以御史之言決一嬋死而欲中起視庶事無使天意以昧欬御史也寒乍暑者非赤乞勤執中賞罰而不罰也鄧保吉有遇於法不當為之言為非赤乞勤執中賞罰而不罰也鄧保吉有遇於法不當為

兩者政事未決也陛執中為樞宋病而家居者百日矣陛下以御史之言為非赤乍寒乍暑者未當賞而賞當罰而不罰也

AN UNEXPECTED GUEST | From the pages of *Lidai mingchen zouyi*, in which the "guest star" of 1054 is described.

Nearly a millennium ago, in the year 1054, astronomers around the world recorded the sudden appearance of a new bright star. It was so bright, it could be seen in the daytime. Chinese astronomers called the unexpected visitor a "guest star." In a book called *Lidai mingchen zouyi*, they gave this account in 1055: "Your servant considers that, since the fifth month of last year [when] the baleful star appeared, a full year has passed and until now its brilliance has not faded." Located in the Taurus constellation and visible to the naked eye from both the Northern and Southern Hemispheres, the guest star could be seen for two years before it faded from the sky.

Today we know that this peculiar star was a supernova, the powerful explosion of a very massive star at the end of its lifetime. A supernova

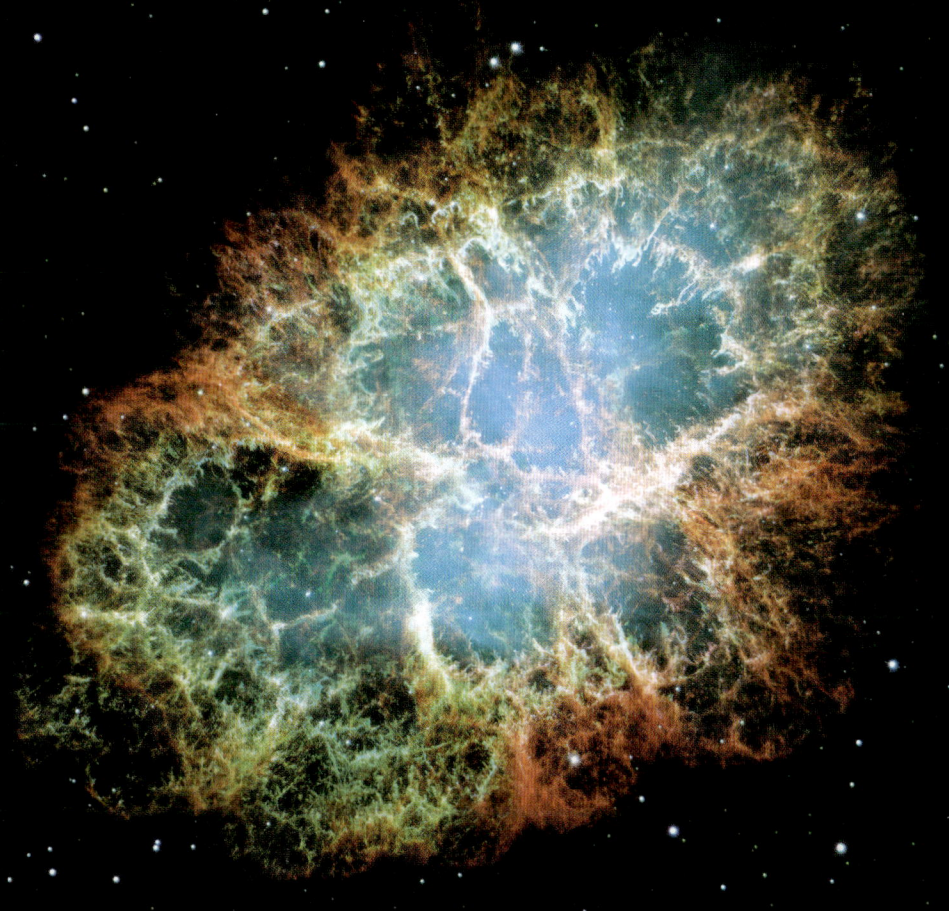

CRAB NEBULA | This Hubble Space Telescope image is the most detailed ever of the Crab Nebula. At roughly 11 light-years across, the shredded remnants of the star consist mostly of hydrogen gas, shown in orange. Other elements expelled during the explosion include oxygen (blue) as well as sulfur (green). The bluish glow emanating from the center comes from light associated with the nebula's rapidly spinning neutron star.

Credit: NASA, ESA and Allison Loll/Jeff Hester (Arizona State University). Acknowledgment: Davide De Martin (ESA/Hubble).

explosion is so energetic, it can momentarily outshine all of the stars in its host galaxy, making it possible to spot them in galaxies far beyond the Milky Way. The explosion itself lasts only tens of seconds, but during this brief time, elements forged inside the star are propelled into interstellar space. The remnant of a supernova will continue expanding for hundreds of years, delivering energy to its surroundings and altering the chemistry of the stellar nurseries that will give birth to the next generation of stars. The star doesn't completely disappear, however; it leaves behind a very compact, rapidly spinning object called a neutron star. The Crab Nebula, the remnant of the supernova observed in 1054, is one of the most well-studied celestial bodies, residing only 6,500 light-years from Earth.

Millions of star deaths like that of the Crab took place in the Milky Way in the years preceding the birth of our Sun. We can be thankful for this. During the death of a star, and only then, are the elements essential for life scattered into the cosmos. We are the offspring of stars.

We Are Connected to Darkness

From end to end, the stellar disk of the Milky Way measures approximately 100,000 light-years. Like the Sun, most stars in the disk are swirling around the Galactic Center. At its heart lies a mammoth black hole, containing as much mass as 4 million Suns. (A black hole is a region of space with gravity so strong, not even light can escape. More about this in Chapter 6.) As its name suggests, a black hole does not emit light, so it can't be observed directly with a telescope. So how do we know that black holes exist, much less how massive they are?

It turns out that black holes like to eat anything that gets too close. And black holes are messy eaters. The supermassive black hole lying at the center of our Galaxy, called Sagittarius A*,* induces stars and gas clouds in its

---

* This is pronounced as "Sagittarius *A-star*." Astronomers typically shorten the name to Sgr A*, pronounced "sadge-ay-star."

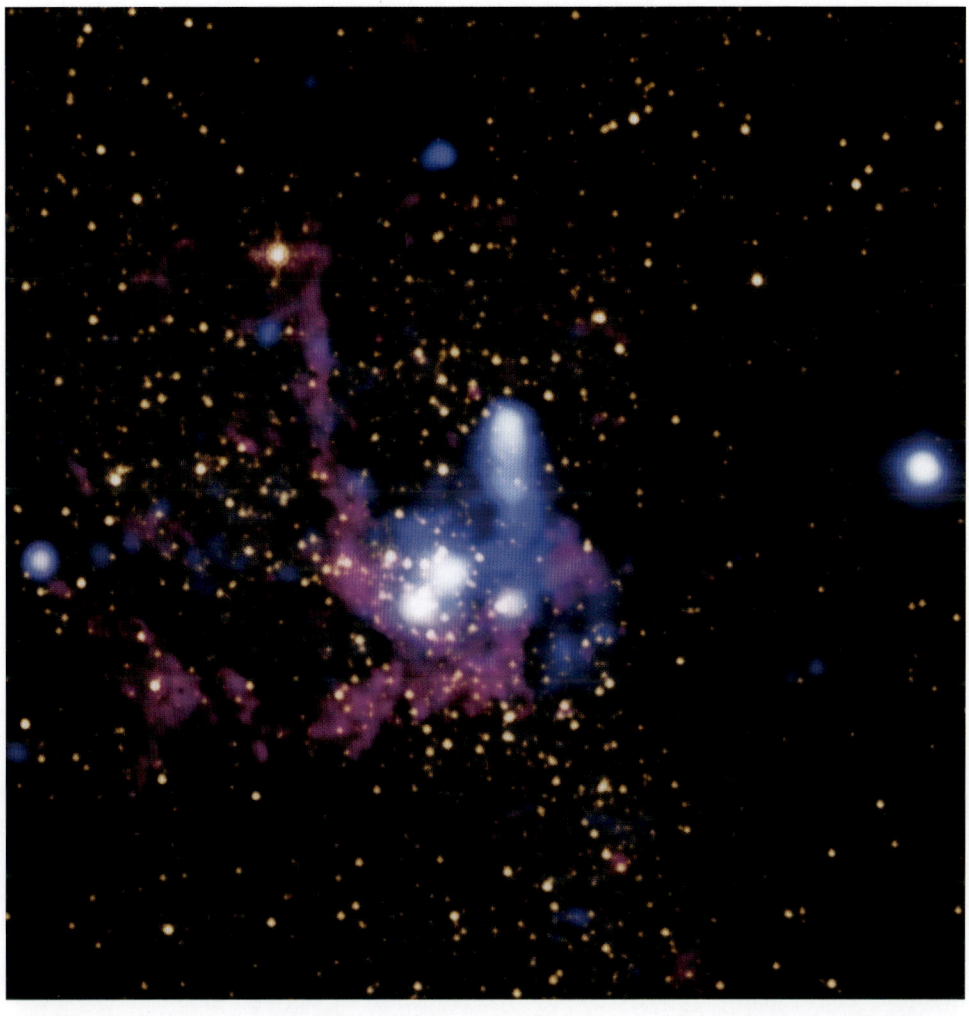

SAGITTARIUS A* | This composite image shows the region around the Milky Way's supermassive black hole, Sagittarius A*. Hot gas expelled by stars near Sagittarius A* is shown in blue. Cool gas and stars are indicated by purple and orange.

vicinity to orbit rapidly around it, due to its colossal gravity. The frenzy of activity taking place just outside Srg A* gives rise to an abundance of light. By measuring the properties of this light, astronomers can explain the motions of the stars and the gas. Then we apply our knowledge of the physics of moving objects to infer the mass of the black hole.

Most massive galaxies tend to have at least one supermassive black hole at the center, but many do not. All galaxies do, however, have this in common: they are gravitationally bound collections of millions, billions, or trillions of stars, with some degree of gas and dust.

Stars are the most striking visual component of galaxies. We often think of them as the main, if not the only, component of our Milky Way. But if we could find a gigantic scale large enough to weigh our Galaxy, we would find that its mass is far greater than the sum of its stars. In fact, less than 5 percent of its total mass can be attributed to stars, planets, or stellar nurseries.

Galaxies are mostly made up of *dark matter*, which is invisible. Astronomers have no idea what dark matter is, but we know that it exists and that it is the most common type of matter in the cosmos. Some 85 percent of all matter in the universe is dark; only 15 percent consists of the "normal" stuff we're familiar with, such as stars, planets, people, and penguins. But although invisible, dark matter is not hidden. We can see its effects. Its enormous gravitational influence causes stars and galaxies to move very rapidly. They'd be slower if dark matter wasn't there.

The physical nature of dark matter remains one of the biggest unsolved mysteries in astronomy. Yet based on indirect evidence, we can infer that it is key to our existence. When a dying star detonates as a supernova, the event would be powerful enough to send most of the material from the exploding star flying out of the Milky Way if the Galaxy's gravity wasn't strong enough to retain it. Dark matter is the biggest source of gravity in all galaxies. Because of it, the elements necessary for life aren't completely blown into intergalactic space. Instead, they can stick around to become incorporated into subsequent generations of stars and their planets. It's because of dark matter that the Milky Way was able to form in the first place.

We Are Connected to the Cosmos

Let's now imagine stepping outside the Milky Way. We find that our Galaxy is one in a swarm of dozens, though most are smaller in size and less massive. The Milky Way is the second-biggest galaxy in our Local Group, with the first-place honor going to Andromeda, a spiral galaxy with an estimated one *trillion* stars, around 400 times as many stars as the Milky Way has! Andromeda is visible from the Northern Hemisphere, and at 2.5 million light-years away, it's the most distant celestial object that can be seen with the naked eye.

It's remarkable to think about how the Andromeda we see today is not the Andromeda that exists "now," but the one that existed 2.5 million years ago. On astronomical timescales, this is a cosmic blink of the eye. Nevertheless, it's startling to realize that the light arriving to us from Andromeda during our generation departed when *Australopithecus*, members of the hominin species that Dinkinesh belonged to, were still inhabiting eastern and southern Africa.*

The Local Group is part of the much larger Virgo Cluster, which includes some 1,500 galaxies. And beyond Virgo lie trillions more galaxies. Most belong to groups and clusters that may contain tens of thousands of members.

A famous image taken by the Hubble Space Telescope further deepens our perspective of the universe. This remarkable picture represents only a tiny region of the sky—equivalent to 1/10 the diameter of the full Moon—yet it contains some 10,000 galaxies whose ancient light took billions of years to reach us. Almost every point or blurry patch of light in this picture is a galaxy, each consisting of billions of stars. Most of these galaxies are billions of times fainter than what the unaided eye can see—too faint, even, to be imaged by the largest ground-based telescopes. Our Solar System hadn't yet been born when the light from many of these galaxies departed on its journey. When

---

* Dinkinesh is the Amharic name for the "Lucy" specimen discovered in the Awash Valley of Ethiopia in 1974. Dinkinesh means "the marvelous one."

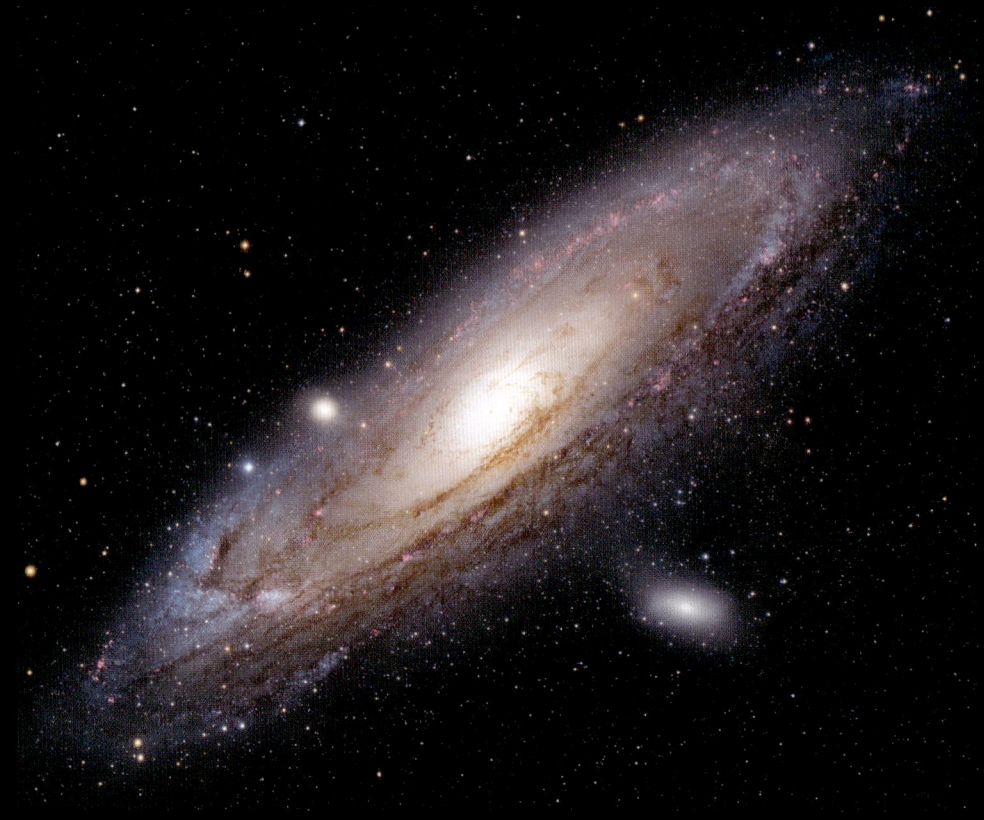

THE ANDROMEDA GALAXY | The largest galaxy in the Local Group, Andromeda is named after an Ethiopian princess in Greek mythology.

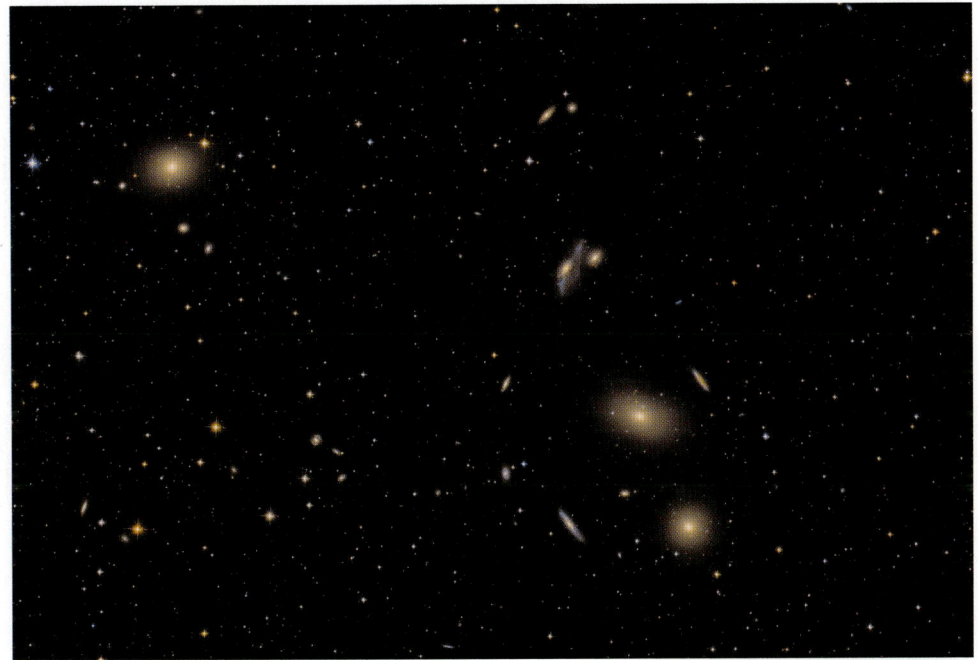

THE VIRGO CLUSTER | About 50 million light-years away, the Virgo Cluster contains more than 1,000 galaxies of many different sizes and shapes.

we look at this picture, we're *time traveling to the past,* when the universe was in its infancy.

Measurements of the light from these galaxies tell us something about their movement. They're not static. In fact, they're speeding away from us at tens of thousands of miles per hour. Galaxies farther away are receding faster, in every direction of the sky. Everywhere we look, we find distant groups of galaxies receding at speeds that increase in proportion to their distance, a phenomenon first observed in 1929 by Edwin Hubble. The best possible explanation for what's happening is this: *the universe is expanding.*

The notion of a growing universe may seem rather mind-bending. But here's another stretch for the imagination. Though it might seem like it, we're not at the center of the expansion. From our vantage point in the Milky Way, we observe distant galaxies flying away from us in every direction. Yet our experience is not unique. An alien astronomer living on a planet in some

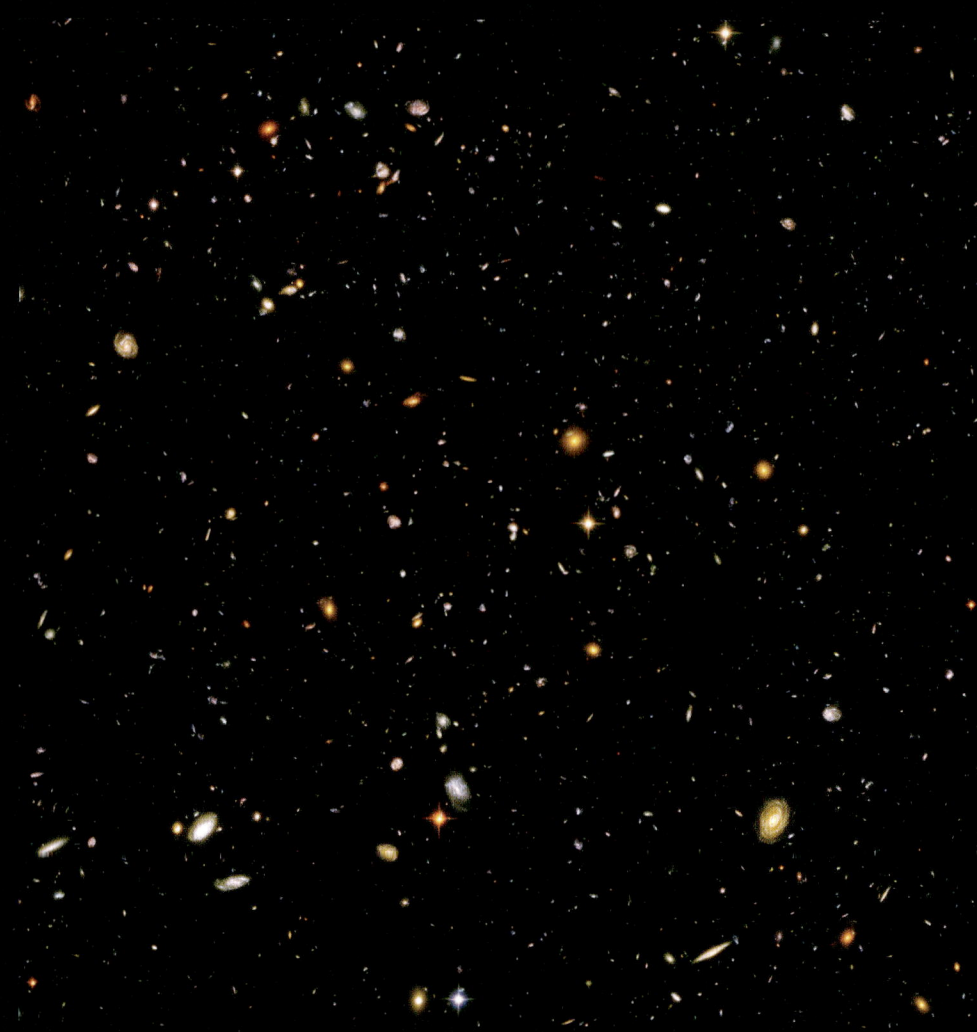

HUBBLE ULTRA DEEP FIELD | Thousands of ancient galaxies imaged by the Hubble Space Telescope. *Credit: NASA, ESA, and S. Beckwith (STScI) and the HUDF Team.*

other distant galaxy would also observe most galaxies flying away from it. How is this possible? *There is no center* to the expansion of the universe.

What's more, the universe is not expanding *into* anything. Space itself is expanding. And it's doing so at an increasing rate. For the past nine billion years of its history, the universe's growth has been accelerating.

Since the universe is growing, this means that everything was closer together in the past. If we rewind/turn back the cosmic clock about 13.8 billion years, we come to a time when the universe was an almost infinitely small point, unimaginably dense, and inconceivably hot. This was the beginning of time, the start of the expansion of space. We call it the *Big Bang*.

Civilization as we know it began to emerge some six thousand years ago, with the invention of writing and the beginning of recorded human history. Modern humans first appeared about 300,000 years before that. Our earliest ancestor, Dinkinesh, "the marvelous one," was born 3.2 million years ago. However you look at it, compared to the universe's 13.8-billion-year history, humanity is a mere blip on the cosmological clock.

As young as humanity is, we are connected with the ancient universe in very real ways. The atoms in our bodies are billions of years old, at least as old as the Solar System. Roughly 1/10 of the mass of the human body is composed of hydrogen, the simplest and most common element in the universe. Hydrogen is also the oldest element, most of it having been formed in the Big Bang. The hydrogen atoms composing the organic molecules and water in your body, including the blood flowing through your veins, are nearly 13.8 billion years old. You are literally an embodiment of the ancient universe.

We Are Connected to the Future

When I began my journey in astronomy, I found it thrilling, yet also unsettling, to discover how vast, old, dynamic, mysterious, and powerful the universe is. Textbooks and scientific articles painted a picture of a cosmos at once remote, indifferent, violent, and foreign. Humanity, by comparison, was insignificant, adrift. Within this frightening framework, it isn't hard to slip from a healthy humility in the face of the cosmos's many wonders to a sense of complete alienation.

I wasn't satisfied with this story. The way science is practiced in most institutions today—with its machinelike picture of the universe—is dominated by a Eurocentric worldview. As I studied, I also quickly grew tired of the overrepresentation of white men in classrooms and textbooks. I felt compelled to relate to science—and to the universe itself—in a way that celebrates the stories of all human beings. Viewing the cosmos through the lens of other cultures—and through art—expanded my perspective. It deepened my appreciation for the preciousness of human life and confirmed my intuition of how intimately interconnected we are.

Science is as culturally rooted as art. Its practice and its products emanate from individuals and communities imbued with the values, motivations, and beliefs of the dominant culture in which they are situated. How science develops in a society is as much a matter of culture as it is of history.

What would happen if, instead of imagining the universe as a cold, lifeless, foreign place, we thought of it as home? If we reminded ourselves that seemingly remote phenomena such as exploding stars and dark matter played an intimate role in our coming into existence? What if we told the story of a cosmos in which, in the words of the Native American leader Chief Seattle, "All things are connected, like the blood that unites us all"?

Our stories aren't just about the past; they point toward our future. Like the light shining from a star, stories are generative, full of creative power. Stories are the fuel for our imagination, that magical spark that can propel our dreams into reality.

CHAPTER THREE

LIGHT—THE BREATH OF COLOR

Colors are light's suffering and joy.
— Johann Wolfgang von Goethe —

Color says something, directly or metaphorically.
— Toni Morrison —

A single ray of light from a distant star falling upon the eye of a tyrant
in bygone times may have altered the course of his life, may have changed
the destiny of nations, may have transformed the surface of the globe, so
intricate, so inconceivably complex are the processes in Nature.
— Nikola Tesla —

E verything in the universe is in motion, including light. Dynamic, per-petually moving light surrounds us all the time. Even in the dark we are immersed in a sea of invisible radiation generated by our technology and by natural processes, both terrestrial and celestial. Light is everywhere around us and also quite literally inside us—because anything that's in motion generates light.

This little light of mine, I'm gonna let it shine.
This little light of mine, I'm gonna let it shine.
This little light of mine, I'm gonna let it shine.
Let it shine, let it shine, let it shine.

The old gospel song draws upon the idea that light is at the heart of who we are. We are able to shine, to live and respond from the heart. Another familiar idea: *Truth is light.* Truth illuminates, gives new understanding. Truth sometimes can hurt, as when too much of it arrives at once, like the shock of looking directly at the Sun. Truth can be reflected, and it can also be distorted. It can be concealed—and it can burst out of hiding. Truth moves us, pushing us to greater awareness. It moves like light.

But what exactly *is* light? How does it enable us to see what we see? Why do leaves look green and the sky blue? When we stand before a moonlit lake or a spectacular sunrise, light can appear ethereal, magical. But light is entirely real, and its properties can be understood physically.

Our modern understanding of light goes back to Ibn al-Haytham, a 10th- and 11th-century Muslim polymath born in Iraq who spent much of his life in Egypt. Widely considered the first theoretical physicist, Ibn al-Haytham was also an early proponent of the modern scientific method. He tested his hypotheses by conducting experiments that could be independently reproduced. By doing this, he disproved earlier theories concerning the nature of light.

In ancient Greece, scholars from Plato and Euclid to Ptolemy argued that human sight works by means of light emanating from the eye. The theory was initially put forth by Empedocles, a Greek philosopher of the 5th century BCE who compared the eyes to lanterns: The light streaming from them and touching objects allows us to see those objects. Critics poked holes in this scheme, pointing out that if this were true, we should be able to see clearly in the dark. Plato argued that different kinds of light existed. *External*

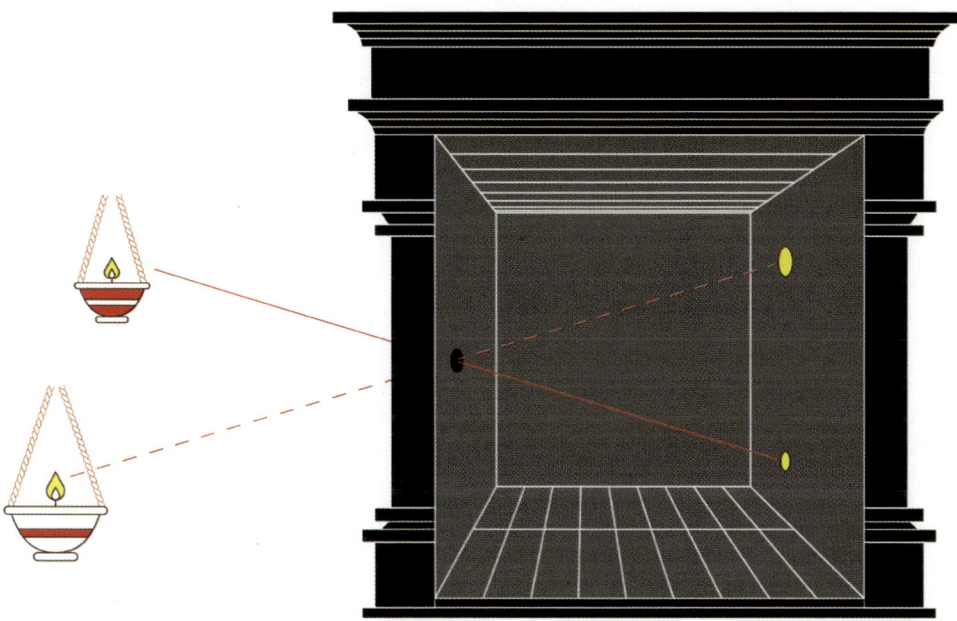

IBN AL-HAYTHAM'S "DARK ROOM" EXPERIMENT | Ibn al-Haytham demonstrated that light rays travel in straight lines.

light—from the Sun, for instance—was needed to complement the *internal* light of the eyes. Others propounded alternative theories of vision, and there's evidence that thinkers in other parts of the ancient world had developed more sophisticated models. But for at least a thousand years, many people believed the eye was a type of lantern.

In the early 11th century, Ibn al-Haytham proved this theory wrong by conducting a simple experiment. He stood in a darkened room with a small hole in one wall. Outside the hole two lanterns were placed at different heights. Inside the room, Ibn al-Haytham then observed two bright spots at different locations on the opposite wall, each in a direct line from the hole to its respective lantern. When one of the lanterns was covered, the corresponding bright spot on the wall disappeared. In this way, Ibn al-Haytham demonstrated two things: light travels in straight lines, and light originates not from the eye but from external sources. He called his brilliant experiment

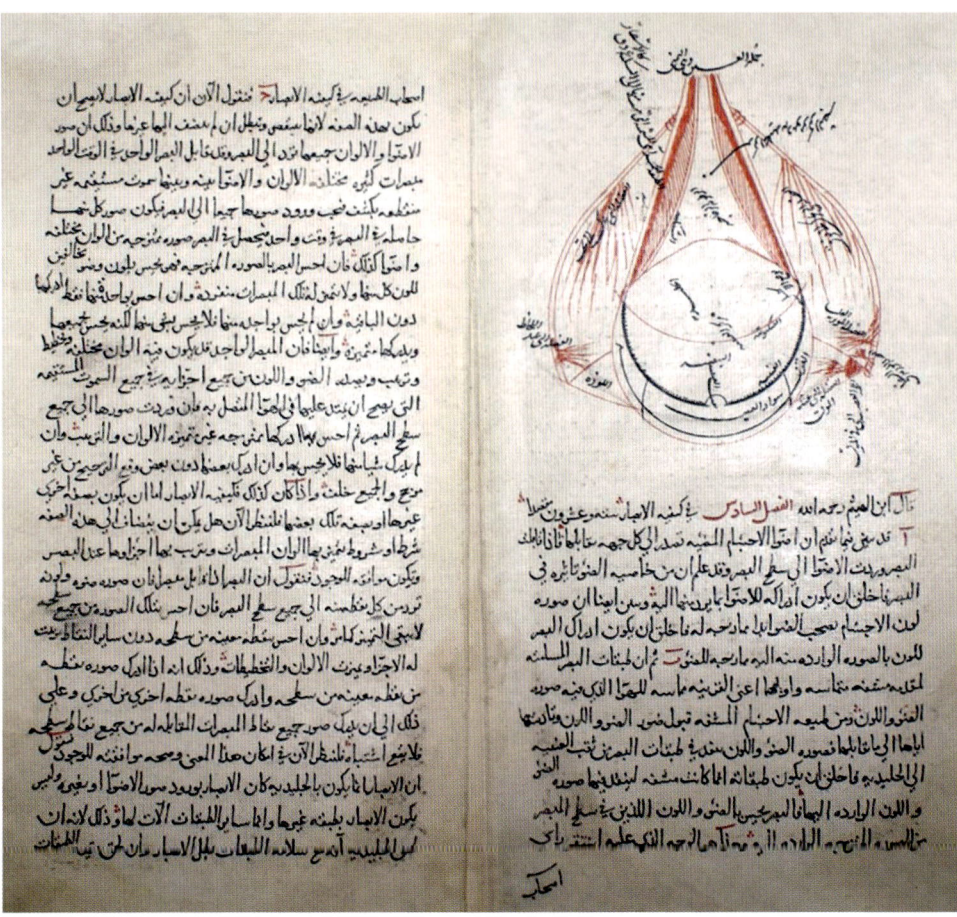

BOOK OF OPTICS | Pages from Ibn al-Haytham's *Book of Optics*, including a diagram of the human eye. *Credit: CPA Media Pte Ltd / Alamy Stock Photo.*

"dark room," or *camera obscura*, the Latin name that Johannes Kepler would later give to Ibn al-Haytham's invention—the pinhole camera.*

Ibn al-Haytham's enormous body of work, including the "dark room" experiment, culminated in his *Book of Optics*, which for several centuries widely influenced thinking about visual perception and the use of perspective in art. Artists and scientists, including Leonardo da Vinci and Galileo, were indebted to him. It would be another 700 years before the next groundbreaking work on the nature of light appeared, with the publication of Isaac Newton's book of a similar name: *Opticks*.

Color Is Light

Today we know that light is energy. When you go outside on a sunny day, you feel warm because the light kissing your skin delivers energy to it. Physically, light can be thought of as both a wave and a particle. A wave is a vibration—a pattern of motion—that carries energy as it moves through space and matter. We're familiar with different types of waves: sound waves, the ripples across the surface of a lake, the waves traveling through the coils of a Slinky, and "the wave" performed by an audience in a stadium. They all have this in common: They need a medium through which to travel. The medium for ripples in a lake is the *water*; the medium for sound waves is *the molecules of air*; the medium for the stadium wave is the fans themselves.

But light is different. It does not need a medium in order to transport energy from Point A to Point B. Light can travel through a vacuum.

We often refer to light as *electromagnetic radiation* or *electromagnetic waves* because it consists of electric fields and magnetic fields vibrating together. An electromagnetic wave can be represented as a series of peaks and valleys.

---

* Johannes Kepler (1571–1630) was a German astronomer who described and published the three laws of planetary motion.

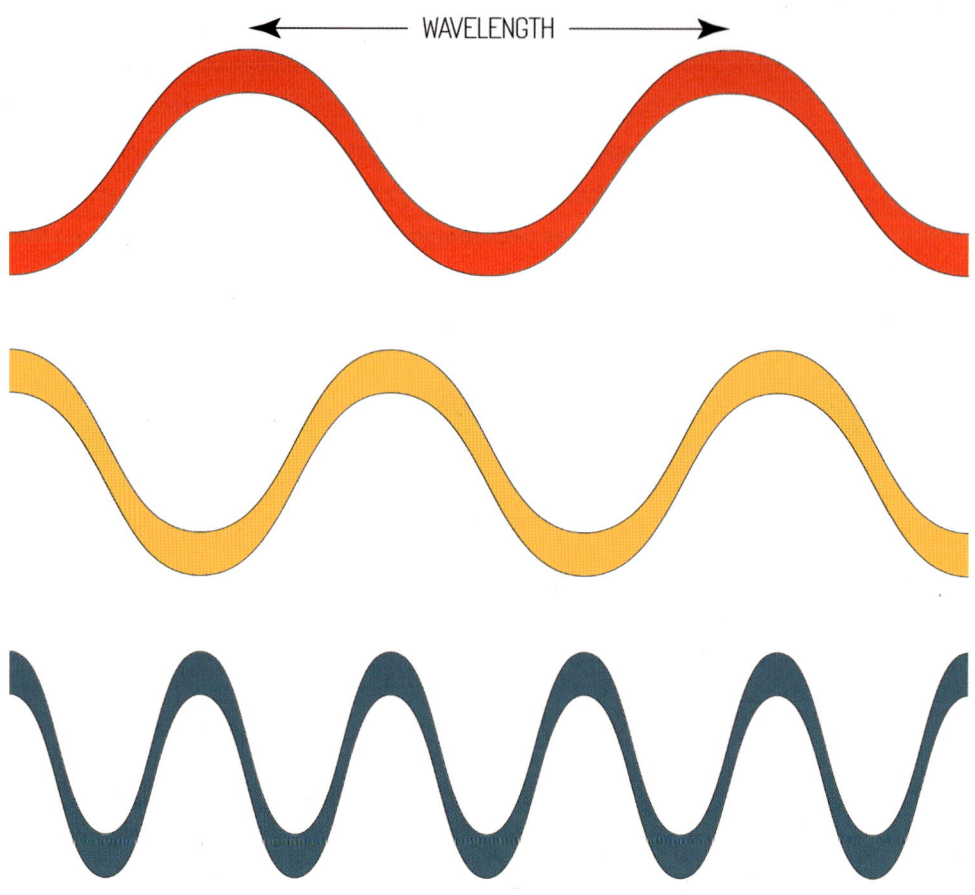

The peaks show the electromagnetic field at its highest level and the valleys at its lowest.

All waves have three fundamental properties—speed, wavelength, and frequency. As already noted, waves move through space—and at certain speeds. The speed of many waves, such as sound waves, depends on the medium through which they are traveling. In air, sound traverses 375 yards in one second. Underwater, it covers the same distance at least four times as fast. Light waves, by comparison, travel at 186,000 miles per second. All the time. No matter what. The speed of light is what we call a *constant of nature*. It is the universal speed limit.

Wavelength is the distance between adjacent peaks in a wave. It can be measured in any unit of length—feet, inches, meters, centimeters. Frequency refers to the number of peaks passing by a fixed point in a set interval of time. These three quantities—light speed, wavelength, and frequency—are interrelated. With light, a wave with a longer wavelength will have a lower frequency than one with a shorter wavelength, for example. But *the speed of light never changes.*

You might have heard that light slows down when it passes through one medium and into a denser one—from air to water, for instance. Here's what actually happens. Water is made of molecules of hydrogen and oxygen (H_2O), which contain electrons. When light enters water, it causes the electrons to move, too, and they generate their own electric waves. These secondary waves combine with the light to produce new waves, which may be bigger or smaller or cancel each other out. The net effect is an *apparent* reduction in the speed of light, though the light waves actually remain unchanged.

In addition to being a wave, light simultaneously exists as a *particle*, called a *photon*. A photon is the smallest possible bundle of electromagnetic energy. Unlike other particles, such as electrons and protons, a photon has no mass and no electrical charge, and it moves at the speed of light.

It can be strange to think of light as simultaneously a wave and a particle, but either description alone is insufficient to explain all the phenomena we observe in nature. The wave nature of light is important to understand because in many situations it behaves like a wave, such as when it appears to slow down in water. Light's particle nature explains how it can travel in straight lines and how we can capture the energy of the Sun using solar panels. Recognition of both aspects of the dual nature of light is needed in order to comprehend the physical world.

The human eye has evolved to see what scientists call *visible light*—the colors of the rainbow. This is only a tiny fraction of all the light that exists. Imagine if light were music. What we'd be able to "hear" of it would correspond to just a few notes, like the melody of "Twinkle, Twinkle, Little Star." We would be deaf to all the light that's out there, which if we could hear it

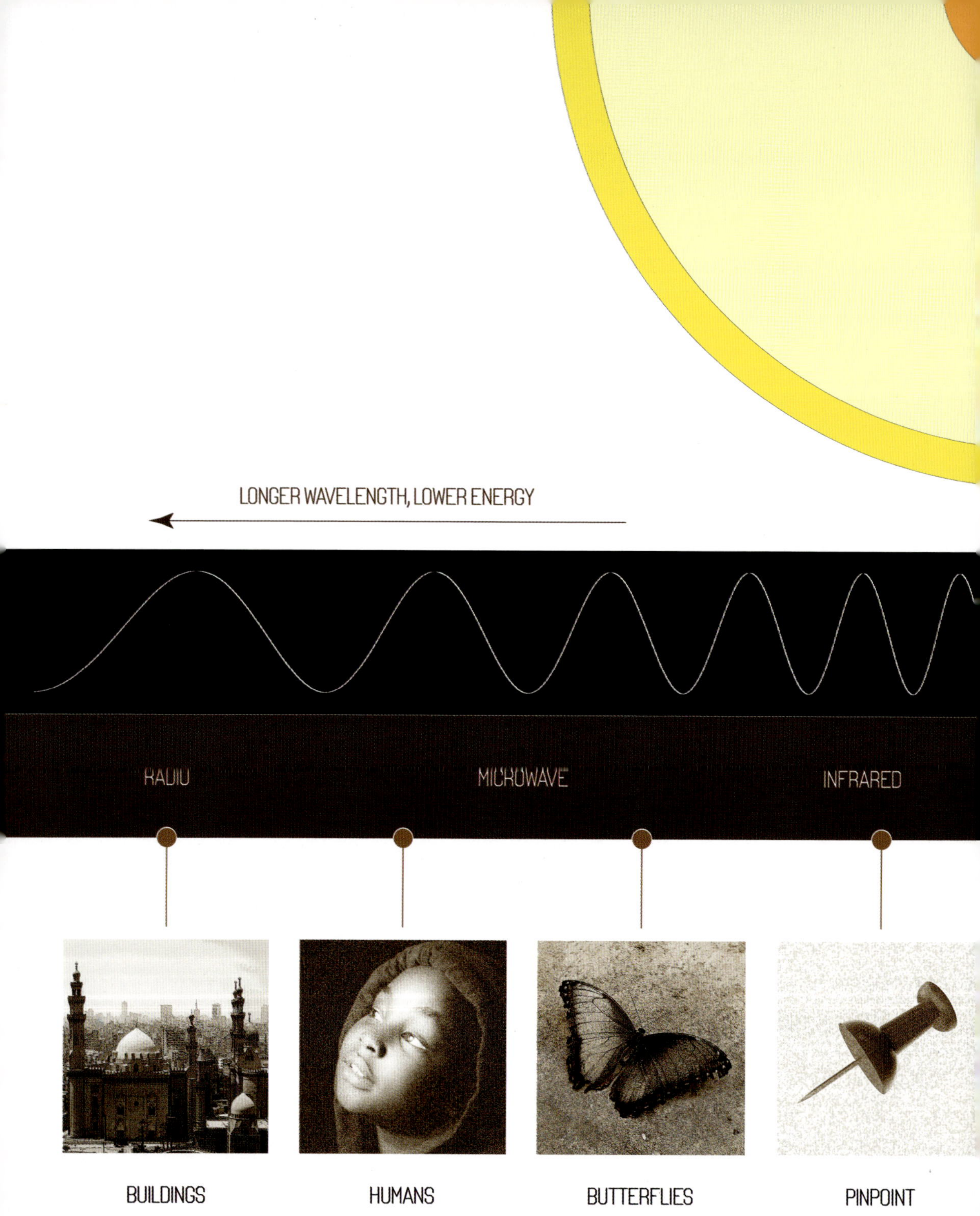

LONGER WAVELENGTH, LOWER ENERGY

RADIO   MICROWAVE   INFRARED

BUILDINGS   HUMANS   BUTTERFLIES   PINPOINT

THE ELECTROMAGNETIC SPECTRUM | The Sun produces light across the electromagnetic spectrum, though most of it is in the form of visible light. This diagram shows the different forms of light, from lowest to highest energies, and compares the approximate scales of their wavelengths with other objects.

SHORTER WAVELENGTH, HIGHER ENERGY →

BLE   ULTRAVIOLET   X-RAY   GAMMA RAY

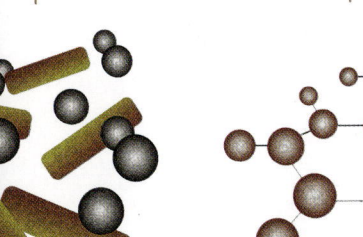

BACTERIA

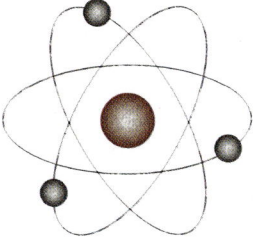

MOLECULES

ATOMS

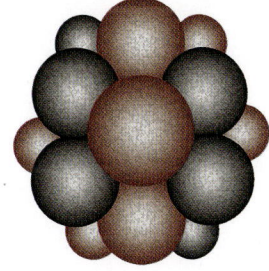

ATOMIC NUCLEI

might sound more like "Another Star," performed by Stevie Wonder and a thousand-piece band.

The term *electromagnetic spectrum* refers to the infinite, continuous range of light, spanning from very short to very long wavelengths. Infrared light, microwaves, and radio waves have longer wavelengths and lower energies; ultraviolet light, x-rays, and gamma rays have shorter wavelengths and much higher energies. All of them are light; all travel at the speed of light.

Visible light represents the very narrow range of wavelengths between about 400 and 700 nanometers.* "Color" is simply the name for the light the human eye can see. *Color is light, and light is color!* The color red is light having wavelengths of close to 700 nanometers, while purple and blue have wavelengths closer to 400 nanometers. The human retina can perceive and distinguish the millions of colors that fall between them.

The Sun generates light at all wavelengths, from gamma rays only hundredths of a nanometer in wavelength to radio waves of just a few millimeters or up to whole miles in length. But much of the light leaving the surface of the Sun, about 44 percent of it, is in the form of visible light, the portion of the electromagnetic spectrum that the human eye can detect. Other animals evolved differently. Butterflies and bees can see ultraviolet light, while some creatures, such as snakes and salmon, can see infrared light.

One of the primary ways that astronomers study stars is to spread their light out into a rainbow, which we call a spectrum.
— Nancy Grace Roman —

Humans *perceive* sunlight as whitish in color. But Isaac Newton showed that when sunlight passes through a prism, the different wavelengths of light fan

---

* To get a sense of how small these numbers are, a nanometer (nm) is a billionth of a meter, and the width of a human hair is around 100,000 nm. Visible light has wavelengths roughly 200 times *smaller* than the thickness of a human hair!

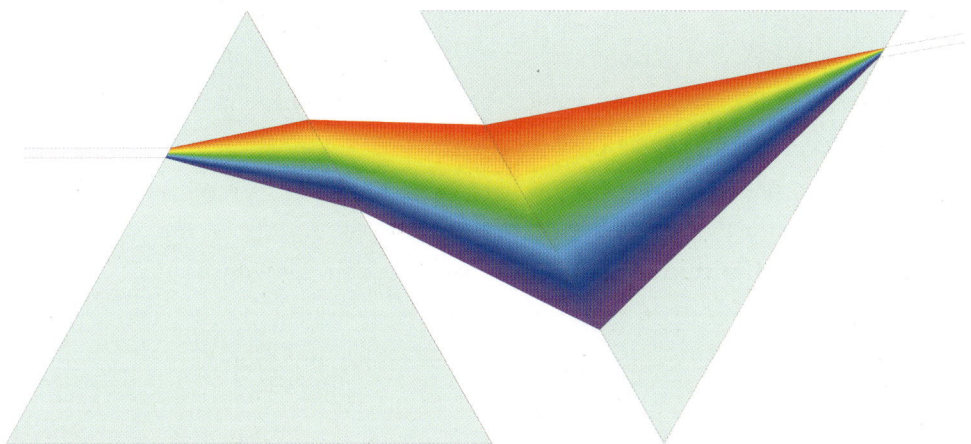

NEWTON'S PRISM EXPERIMENT | White light enters the first prism, refracts into a rainbow, enters the second prism, and rejoins into white light.

apart and become distinct, demonstrating that white light is *not* devoid of color but a combination of all the colors of the rainbow. Newton's famous experiments with prisms led him to conclude that the behavior of light could be explained only by this principle: Light consists of a steady stream of particles. Arguing that the movement of light was in accord with his second law of motion, Newton aimed to discredit an alternative theory proposed by his contemporary Christiaan Huygens, which conceived of light as consisting of waves. Newton placed two prisms at angles to each other. White light shone into the first prism fanned into the rainbow of colors, which then entered the second prism and rejoined into white light. The second prism in this experiment was crucial. It allowed Newton to demonstrate that color was a property not of the glass, as had previously been thought, but of light itself.

Newton's theory of light did not explain everything about its nature, and he got some parts entirely wrong, for example, his notion that light has mass. Physicists at the turn of the 20th century ultimately realized that light has *both* a particle nature and a wave nature.

Photons, themselves invisible and massless, make it possible to see the world around us. Light interacts with matter in a variety of ways. When it shines on an object, it can be *transmitted, absorbed, reflected,* or *refracted.*

Matter can also *emit*, or generate, light. The Sun is the dominant source of light in the Solar System. Like all stars, it is an enormous ball of gas that shines because it generates its own light through nuclear fusion, a process extremely difficult to achieve on Earth.

All sorts of objects on Earth emit light as well, though in ways having nothing to do with fusion. Bioluminescent fish, fireflies, some fungi, and some bacteria, for instance, glow because of biochemical reactions that produce visible photons. Also, anything on Earth that has an average temperature of about 300 Kelvin, nonliving and living—including plants, rocks, animals, and the entire planet itself—gives off infrared light.[*] (At the same time, these objects can *reflect* visible light, as the planets and the Moon reflect sunlight.)

Transmission of light occurs when light passes straight through matter, such as a clear glass window. Cell phones work because they emit and receive low-energy radio waves that can travel long distances, passing through buildings, trees, cars, and people.

Matter can also *absorb* light. On a bright summer's day, we experience this when we feel sunlight warming our skin. Or on a bright winter's day, after a snowfall, when the snow absorbs some sunlight and begins to melt. The absorption of light involves the transfer of light's energy to matter, which heats up as a result and possibly even changes its chemical state.

From the face of a loved one to the face of the Moon, most of what we see on Earth is due to reflected light. Ripe strawberries appear red to us because the molecules making up their flesh reflect red light and absorb the other colors. The feathers of a canary and the sunflowers painted by van Gogh appear yellow because they mirror yellow light and swallow other colors like a sponge. A flower that appears to us as a vibrant violet may take on an entirely different appearance to a butterfly or bee, whose eyes can perceive ultraviolet light. Our atmosphere's ozone layer absorbs most ultraviolet radiation coming

---

[*] The Kelvin is a unit of temperature used by scientists. Three hundred Kelvin is equal to 80 degrees Fahrenheit (27 degrees Celsius).

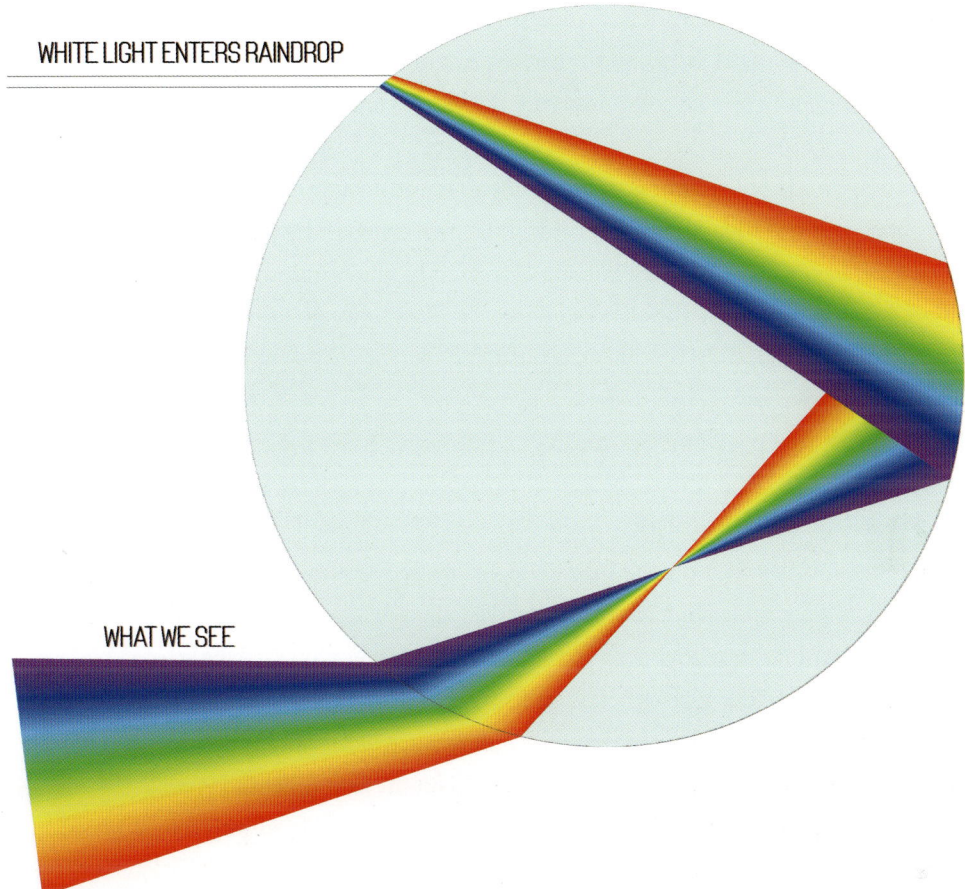

WHITE LIGHT ENTERS RAINDROP

WHAT WE SEE

HOW A RAINBOW FORMS | White light enters a raindrop, refracts, reflects off the back of the raindrop, and continues to refract as it passes through the droplet and emerges. You might notice that the order of colors emerging from the drop is inverted from what you actually see. A fuller explanation of how rainbows work shows that, since each color refracts by a specific angle, only one color from each raindrop makes it to your eyes. Because seeing a rainbow depends on where you are with respect to the Sun, it also turns out that each person sees a different rainbow!

from the Sun, but some of it makes it through and illuminates patterns in nature that we humans are blind to. But these winged creatures can perceive a totally different world of color. Imagine the world through butterfly eyes!

Another thing light can do when it encounters matter is to *refract*, or bend. Refraction occurs when light goes from one medium to another and changes direction, as Ibn Al-Haytham observed in his experiments with glass spheres filled with water. He saw that light was refracted by the water, with different colors being refracted by different amounts. Red and orange light rays were bent a little, and blue light was bent the most. Newton observed the same thing happening in his prism experiments. Anyone who has ever seen a rainbow is familiar with this phenomenon, too. When it's rainy and sunny at the same time, the trillions of raindrops act like tiny prisms that break the sunlight into all of its colors. But the situation is actually more complicated. The raindrops are like prisms *and mirrors* combined. A light ray penetrating a raindrop first bends, then reflects off the back of the droplet, then bends again as it exits—now separated into its component wavelengths. Each color always bends by a certain specific angle, which depends on its wavelength. This is why the colors of the rainbow always appear in the same order. Since light rays reflect off the back of raindrops, you can see a rainbow only when the Sun is behind you.

Imagine sunlight passing through Earth's atmosphere. Of all the colors, blue light tends to get scattered the most by the molecules in the atmosphere. Scattering is a messy form of reflection that occurs when light bounces off a gas, rather than a solid surface. While the other colors of sunlight take a fairly direct path through the atmosphere, blue light is scattered all across the sky. This is why the sky often appears blue.

At sunrise and sunset, the Sun is low on the horizon. Because it is at a lower angle compared to when it is directly overhead around noon, its light has to travel through much more atmosphere before it reaches us at our particular location. Therefore nearly all the blue light gets scattered, leaving behind the redder light that we often see at this time of day. If it happens to be raining, sunlight passing through the raindrops will be filtered of its shorter wavelength colors, and red rainbows can appear!

The Meaning of Color

The way matter interacts with light, reflecting and bending it, absorbing or transmitting it, is an apt metaphor for how humans receive and respond to the influences of their environment. Nina Simone said, "An artist's duty, as far as I'm concerned, is to reflect the times . . . I choose to reflect the times and the situations in which I find myself." In this metaphor, "the times" are equated with light, and the artist can reflect them as the petals of a flower reflect color. To go further with the metaphor, just as a flower reflects certain colors depending on its chemical makeup, how an artist reflects her environment depends on her "chemical makeup." Her identity, beliefs, values, motivations, and feelings—all are ingredients of her mental and emotional "matter" with which external reality (light) can interact.

Expanding on our *truth = light* metaphor, we can say that truth, like sunlight and starlight, is composed of many colors, and it's never revealed to us all at once or in its totality. The truth about a sunset is not merely about the *physical* phenomena involved; a sunset cannot be fully understood solely by explaining that molecules in Earth's atmosphere selectively scatter certain wavelengths of sunlight while allowing others to be transmitted. Likewise, imagine describing one of van Gogh's sunflower paintings to a blind person. How inadequate it would be to say that "yellow" is simply an electromagnetic vibration having a wavelength of approximately 580 nanometers.

Equally true about sunsets and the color yellow are what they make us *feel* and what they signify. The questions posed by science—*why*, *how*, and *what*—lead to only one dimension of what is true. Science can tell us why we see what we see and how the universe works. Through art, we can give meaning to the data and information that science and everyday experience present.

For the artist who loves color, the goal is to understand light so well that it becomes a language she can use to communicate her vision in the most powerful, effective way possible. Thus, she approaches light from the perspectives of color theory, symbolism, its psychological effects, and maybe even its potential for healing.

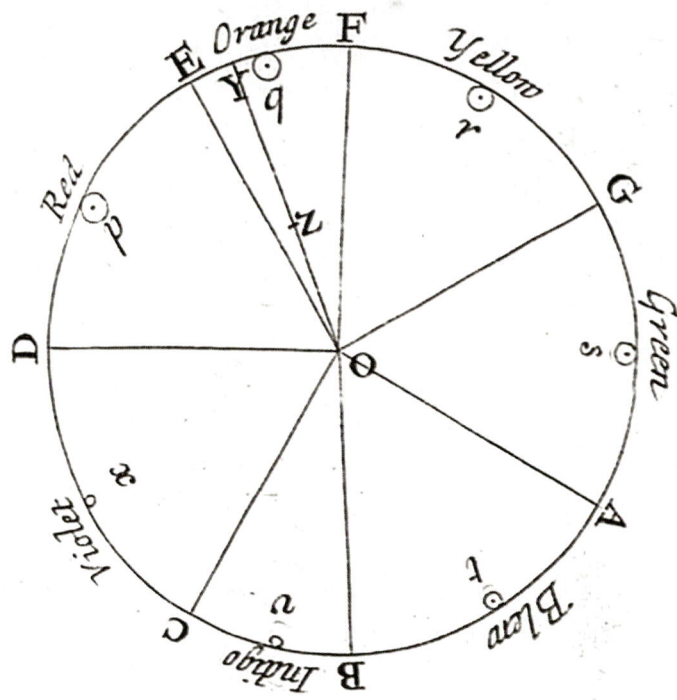

NEWTON'S COLOR WHEEL | Published in *Opticks*, 1704.

Since antiquity, artists, philosophers, and scientists have speculated as to where colors come from, how they relate to one another, and how they can be used to produce certain visual and emotional effects. Over the centuries, in the East and the West, thought and experiment have come together in different theories of color. In the tradition of Western art, before Newton, color was believed to result from the mixing of "light" and "dark." Newton observed that mixing certain colors in the spectrum could produce others, and he joined the ends of the spectrum, forming the first color wheel. Inspired by music and ever the numerologist, Newton mapped his colors—red, orange, yellow, green, blue, indigo, and violet—to the seven musical notes in an octave. Hence, the asymmetric spacing of the circle and the inclusion of the somewhat superfluous sixth color—"indigo."

Some 100 years after the publication of Newton's *Opticks*, the German philosopher and poet Goethe began conducting his own experiments with

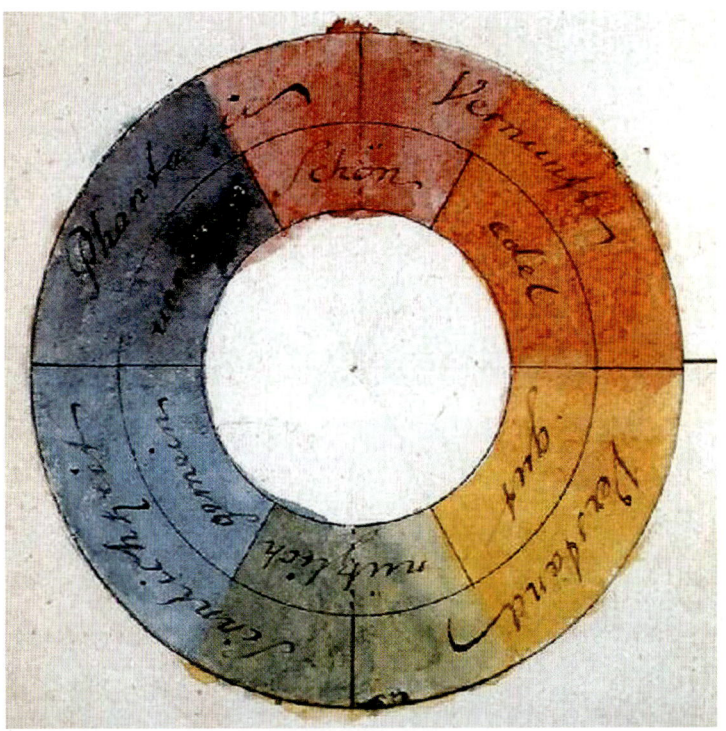

GOETHE'S COLOR WHEEL | Published in *Theory of Colours*, 1810.

color. He was primarily interested in its cognitive and psychological aspects. He created his own color wheel, which bears a much closer resemblance than Newton's to those found in art classrooms today. Now the colors are evenly spaced, so that complementary colors—pairs of hues having the strongest visual contrast—lie directly opposite each other.

Newton's color wheel deals with the relationship between different colors of *light*, while Goethe's concerns the combining of *paints*, or pigments. Combining different colored lights, referred to as additive color mixing, can produce white light. Mixing different-colored pigments, called subtractive mixing, eventually produces black.

Goethe published his color wheel and the results of his numerous experiments in *Theory of Colours* in 1810. Though he gets much of the scientific theory of color wrong, his insights about the impact of color on mood and emotion have had a far-reaching impact on generations of art students:

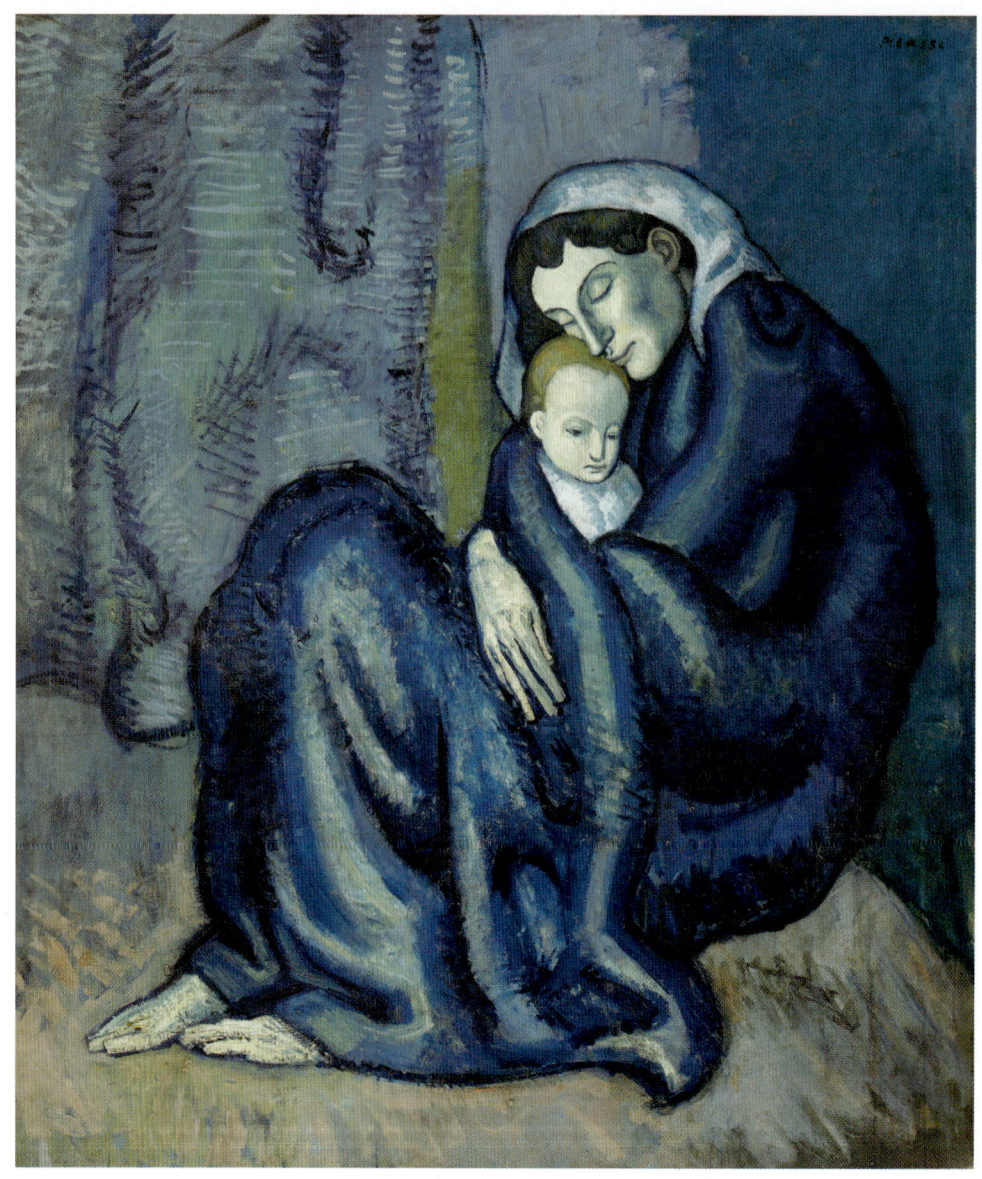

MOTHER AND CHILD | Pablo Picasso, c. 1901, oil on canvas, 44⅛ × 38⅜ in, (112 × 97.5 cm).

© 2024 ESTATE OF PABLO PICASSO / ARTISTS RIGHTS SOCIETY (ARS), NEW YORK.

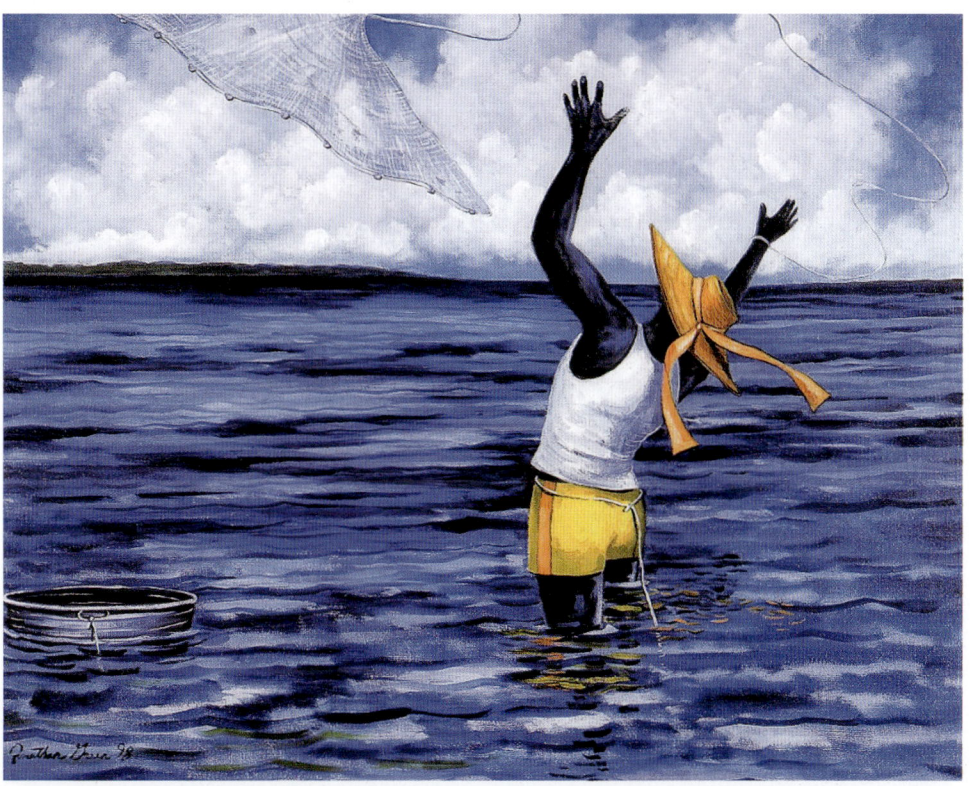

NET CASTING | Jonathan Green, 1998, oil on linen, 18 × 24 in (45.7 × 61 cm).

As yellow is always accompanied with light, so it may be said that blue still brings a principle of darkness with it.

This colour has a peculiar and almost indescribable effect on the eye. As a hue it is powerful—but it is on the negative side, and in its highest purity is, as it were, a stimulating negation. Its appearance, then, is a kind of contradiction between excitement and repose.

As the upper sky and distant mountains appear blue, so a blue surface seems to retire from us.

But as we readily follow an agreeable object that flies from us, so we love to contemplate blue—not because it advances to us, but because it draws us after it.

Blue gives us an impression of cold, and thus, again, reminds us of shade. We have before spoken of its affinity with black.

Reading Goethe's description of the color blue, I cannot help but recall Picasso's painting *Mother and Child* of 1901, inspired by his visit to the women's prison in Saint-Lazare. I get that "impression of cold." Then I think of Jonathan Green's use of blue, and I feel neither cold nor shade. Quite the opposite. In Green's painting *Net Casting*, the vibrant cobalt and ultramarine blue of the sea are warm, luminous, inviting. I feel tempted to step in with the fisherman, who stands there in shorts with water up to his knees, whose dark skin reflects the warm blue light of the hot Sun.

Light is something that cannot be reproduced, but must
be represented by something else—by color.
— *Paul Cézanne* —

In the early 1900s, the artist and inventor Albert Munsell introduced a significant innovation in color theory. He was interested in classifying color in a way that would help us better communicate with it. Initially, he tried to organize his system on a sphere, building on the work of artists who had developed 3D models of the color spectrum. Realizing that the relationships between colors were too complicated to be represented this way, he created an elaborate "color tree" based on three dimensions: *hue* (that is, color—blue, yellow, and so on), *value* (how light or dark), and *chroma* (the purity or saturation of a color). Munsell created hundreds of multicolored "leaves" for his 3D tree, using his patented photometer, an instrument that measures the intensity of light, to measure the hundreds of scrupulously selected colors.

Munsell's system, published in his book *A Color Notation*, was considered a triumph of the intersection of art and science; his model of the three dimensions of color continues to be taught in art schools. Munsell made another important contribution to color theory by criticizing the traditional notion that red, yellow, and blue are the "primary colors." Many of us remember being taught that combining these "pure colors" in certain ways can produce all the others on the spectrum. But it turns out that any three colors can be mixed to create a subset of the spectrum. Certain trios can produce a wider range of colors, but there's nothing intrinsically "primary" about red, yellow, and blue. No fixed set of pigments can produce *all* the colors the human eye can perceive.

Moreover, the idea of what colors are "primary" changes with cultural context. In traditional Chinese and Japanese culture, white, black, red, yellow, and blue/green are the principal hues, emblematic of natural phenomena and certain emotions or values (such as water, fire, joy, purity, harmony, growth). In the arts of many African peoples, red, white, and black are primary, having specific symbolic significance that varies from culture to culture. In different parts of the world, a single color can have strikingly different associations. In the modern West, for instance, black is the color of mourning and often carries negative connotations. For some West African cultures, black represents rebirth and maturity, whereas white may symbolize the world of the ancestors.

Instead of trying to come up with a universal theory of color, the German-born artist Josef Albers was interested in its experiential possibilities. Albers called color "the most relative medium in art," stating that "one and the same color evokes innumerable readings." He conducted brilliant experiments that explored the dynamic *relationships* between colors, recognizing that the perception of a particular color depends on the colors around it. In one of his famous demonstrations, Albers placed two little same-colored squares on different-colored backgrounds. Most people see the squares as different in color, but they're actually identical.

Whereas scientists agree on the physical nature of light, there's no single definitive theory of color. Physically, light behaves in predictable ways. But our actual *experience* of color as human beings is a complex, subjective phenomenon. How we think about and respond to color depends on numerous interacting factors, including age, personal experience, mental and emotional state, and culture. Yet theories of color have their uses. Visual artists study and apply them to learn how to work effectively with pigments and how to achieve harmony, proportion, and movement in their art. But the artist knows that rules are made to be broken, that in order to create something truly special, the magic and mystery of the human personality must at some point enter the equation.

Color Says Something

I could stare forever at the exquisitely black face of the woman in Romare Bearden's *Falling Star*. In this painting, the traditional rules of color theory give way to the artist's powerful vision. There's *feeling* in these colors that theory can't prescribe. Bearden does not completely abandon well-established traditions; instead he revitalizes and builds upon them.

Bearden was born in Charlotte, North Carolina, in 1911. He majored in mathematics in college and spent three years in medical school before putting aside the stethoscope for paintbrushes and, eventually, collage. A master colorist, he took the colors and images so familiar to us and weaved entirely new stories from them. His work is a lens through which we can see the world in a new way.

In Bearden's world of *Falling Star*, coral-colored floors match coral arms and coral lamps; a beam of light casts an unnatural, multicolored shadow, then cuts into the fabric of the woman's dress; a cool white bedspread, with its precise floral design, flawlessly juxtaposes the vibrant, geometric blocks of warm color. Feelings of enchantment, curiosity, and tenderness are aroused. You're surprised by affection, too, and you feel the desire to laugh. The woman serenely blows her tea, her daintily raised pinky pointing, unbeknownst to her, to a meteoroid streaking crazily through the atmosphere, just outside her window. You're reminded of the often comic absurdity of life, of the certainty that beautiful surprises lie in wait around every corner.

And then there's Bearden's arresting use of color in *Circe*, his reimagining of the banished goddess who seduced Odysseus and turned his men into swine. She appears three times in his series of 20 collages based on incidents

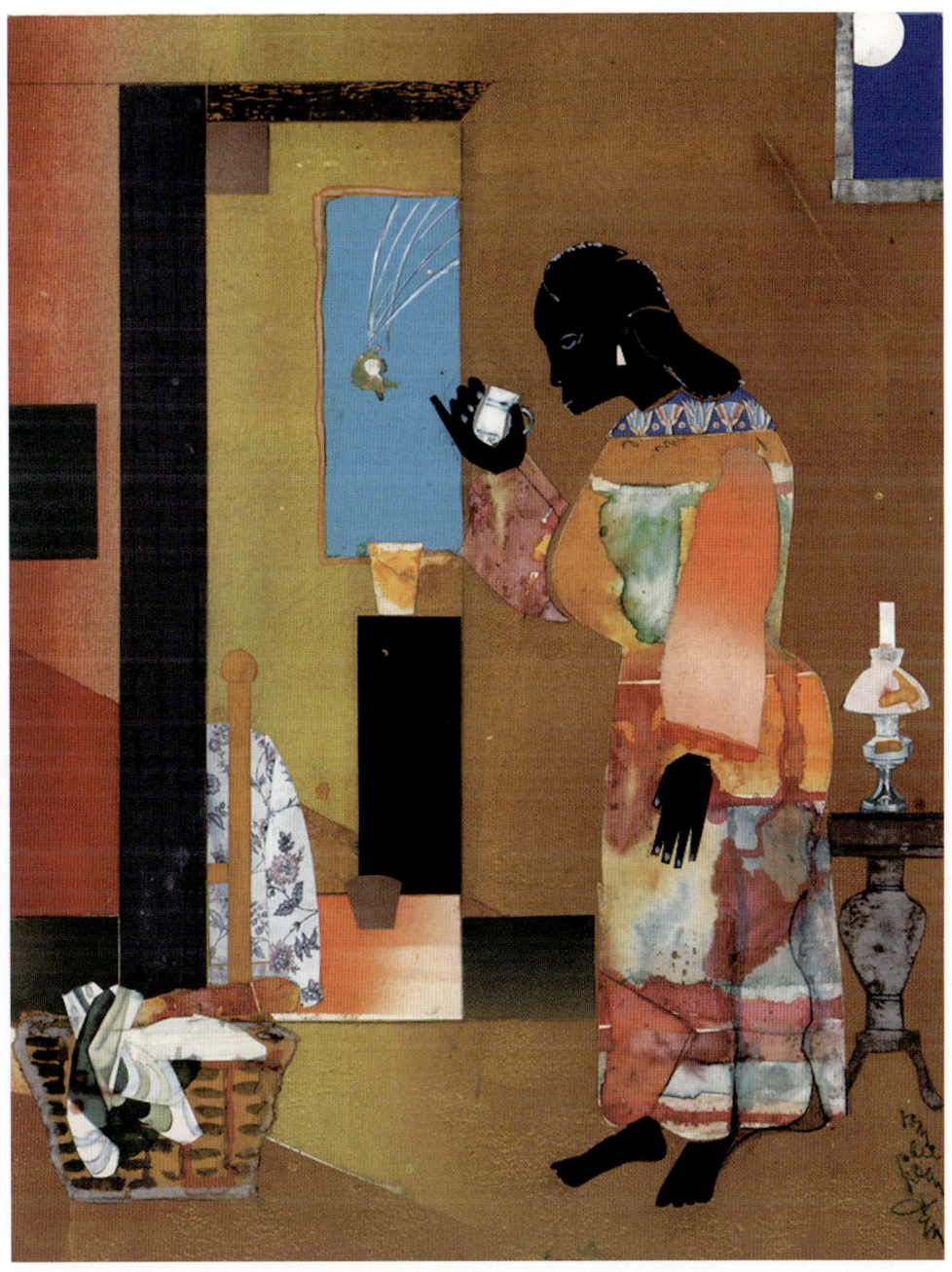

FALLING STAR | Romare Bearden, 1979, collage, 14 × 18 in (35.6 × 45.7 cm).

© 2024 ROMARE BEARDEN FOUNDATION / LICENSED BY VAGA AT ARTISTS RIGHTS SOCIETY (ARS), NEW YORK.

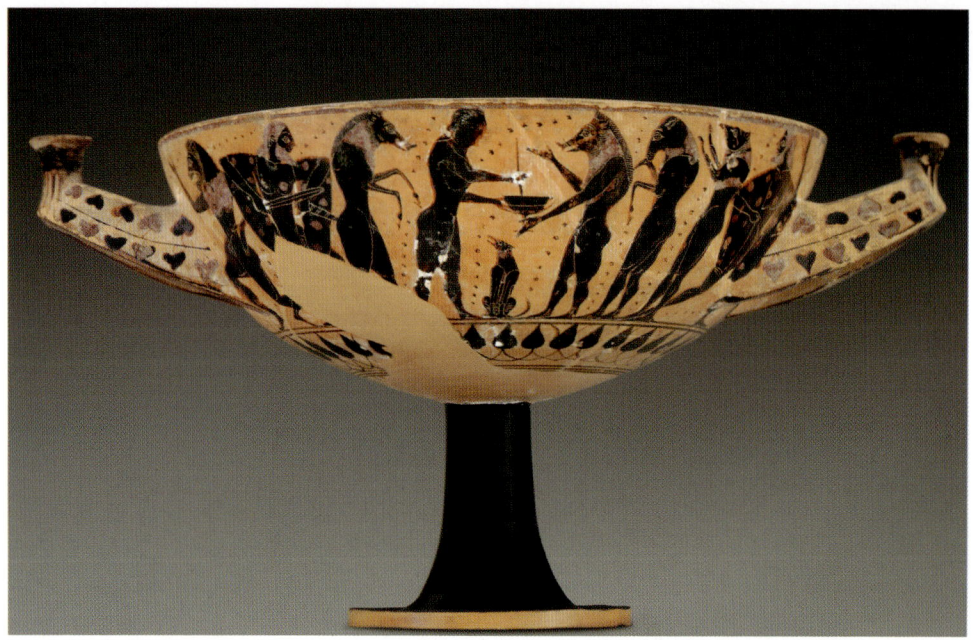

CIRCE | Circe is depicted here in a scene from *The Odyssey* on a kylix, an ancient Greek drinking cup (about 550–525 BCE). *Photograph © Museum of Fine Arts, Boston.*

from Homer's *Odyssey*. Here, Bearden dialogues and expands upon the classical Greek poem, considered to be a foundation of Western literature. The story recounts the hero Odysseus's long, eventful journey home after the Trojan War; he is constantly hindered by various gods and goddesses along the way. Bearden reinterprets the story as a myth set in Africa, with Circe as a black woman. Like all of the characters depicted in *A Black Odyssey*, she's as black as possible. She's some kind of high priestess or royal figure; she's powerful, graceful, and intelligent; her beauty captivates.

My favorite Circe is the one with the blue snake winding about her arm; a screaming red backdrop accentuates her blackness—a rich, seductive black unlike any black you've ever seen. The dramatic tension introduced by the red is heightened by that color's contrast to the muted grays of the background column and the foreground post. Interestingly, Bearden achieved his striking effects of color by limiting his palette: "I found that even in spite of the fact that I have to restrict my color, that just using a few colors can give me quite

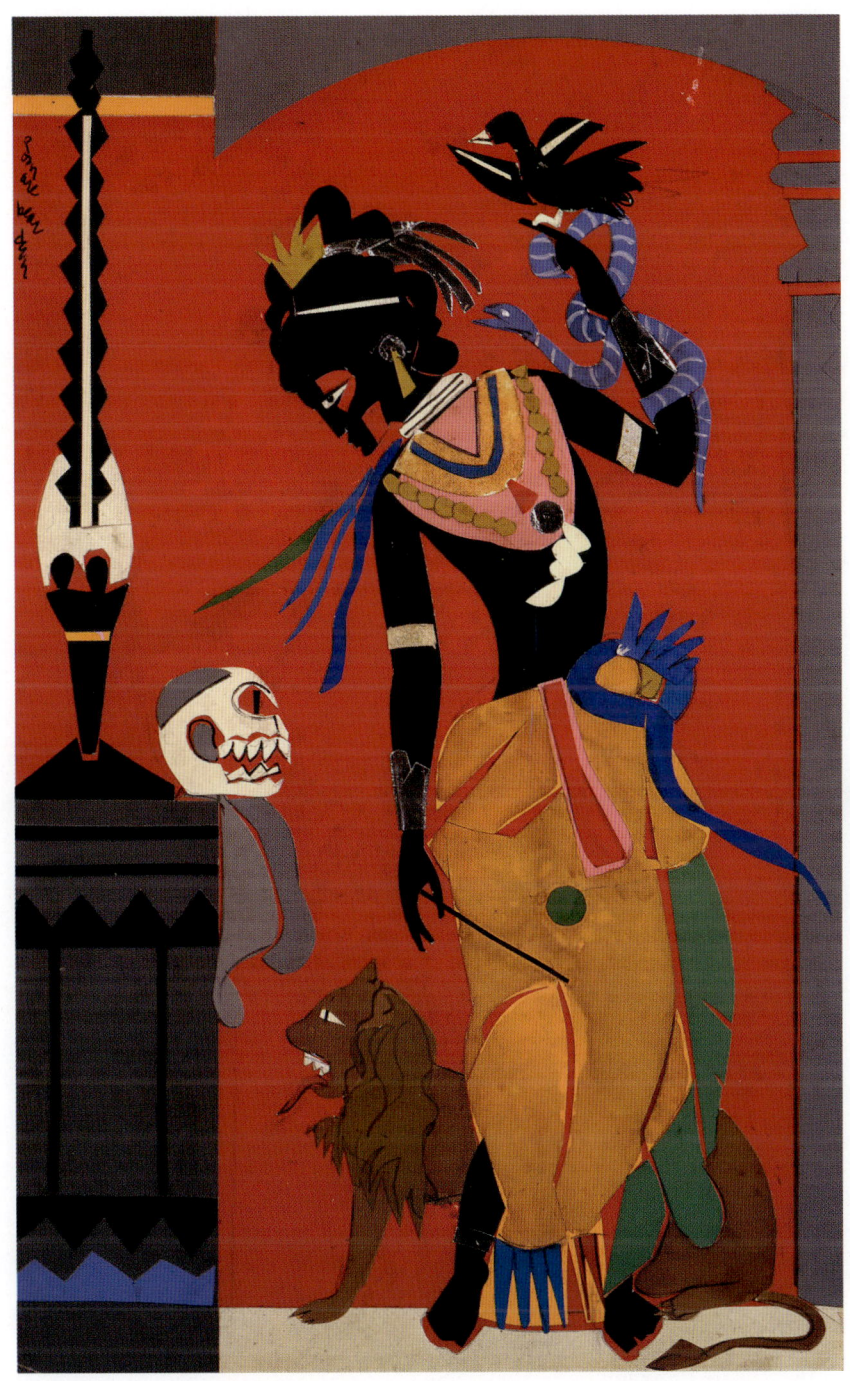

UNTITLED ("BLACK CIRCE") | Romare Bearden, 1977, collage on board, 15 × 9⅜ in (38.1 × 23.8 cm). © 2024 ROMARE BEARDEN FOUNDATION / LICENSED BY VAGA AT ARTISTS RIGHTS SOCIETY (ARS), NEW YORK.

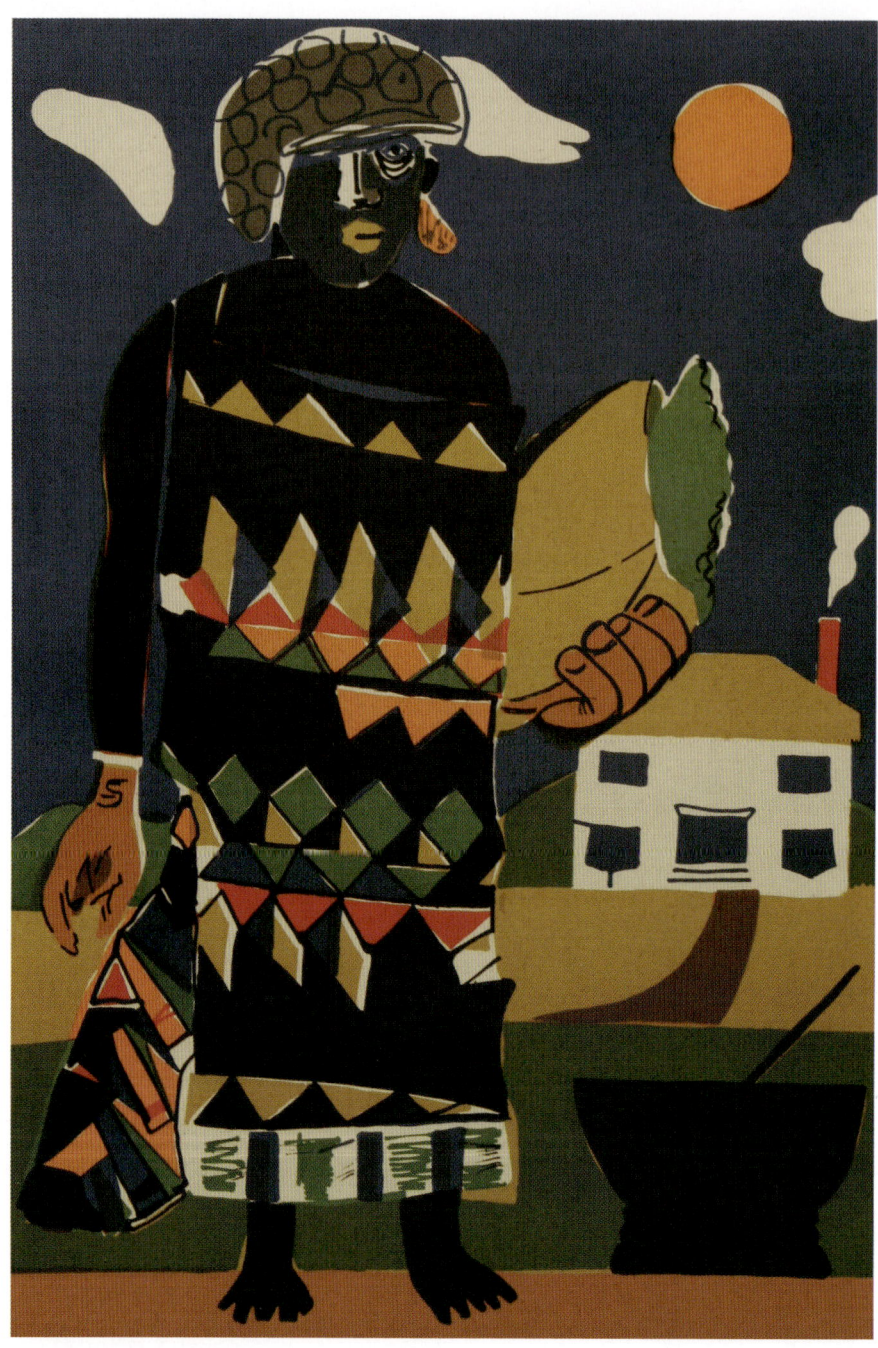

PILATE | Romare Bearden, 1979, lithograph, ed. 7/25, 22½ × 15¼ in (57.1 × 38.7 cm). Bearden gifted the original of this lithograph to Toni Morrison.

a range . . . what I try to do is relate all my colors to gray and then put in in a few places a few dissonant accents . . . what stands up the longest almost are paintings where you feel the absence of color. Like the Chinese paintings, done with just washes of gray or a very little touch of color."

Bearden maintained that art derives from other art, that the art of the world forms part of our common heritage. Rooted in the culturally specific—"the black American experience"—his work incorporates layers of reference to the art and stories of other cultures, giving it depth and extension. "As for my own work," he once said, "what I've tried to do is to take the elements of Afro American life in this country as I see it, that is to place it in a universal framework." The *Odyssey* series is influenced not only by Western European art but African and Asian art. Bearden is engaged in more than tepid multiculturalism. By fusing elements from a variety of traditions, Bearden powerfully illuminates our common humanity.

Toni Morrison also took up Circe as a character in her novel *Song of Solomon*, published in 1977, the same year that Bearden created the *Odyssey* series. Morrison's Circe is also a black woman. She is a nearly 100-year-old midwife who, similar to Homer's Circe, profoundly influences the journey of the novel's archetypal hero, Milkman. The lush blackness of Bearden's Circe calls to mind the words of Milkman's protector, Pilate—a very black woman herself—regarding the multiplicity of the color black:

> You think dark is just one color, but it ain't. There're five or six kinds of black. Some silky, some woolly. Some just empty. Some like fingers. And it don't stay still. It moves and changes from one kind of black to another. Saying something is pitch black is like saying something is green. What kind of green? Green like my bottles? Green like a grasshopper? Green like a cucumber, lettuce, or green like the sky is just before it breaks loose to storm? Well, night black is the same way. May as well be a rainbow.

Morrison, like Bearden, challenges racist ways of interpreting the color black in the Western tradition. She reappropriates a symbol of the culture that has been used as a tool of degradation and restores it as an instrument of creative power.

Color is an important element in *Song of Solomon* and other of Morrison's novels. Describing her creative process, she says, "I need three kinds of information to complete, sometimes even start, a narrative. Once I've settled on an idea and the story through which to examine it, I need the *structure*, the *sound*, the *palette*—not necessarily in that order." Morrison goes on to say, "The third, palette, color, is one of the last and most crucial of my decisions in developing a text. I don't use color to 'prettify' or please, or provide atmospherics, but to imply and delineate the themes within the narrative. Color *says* something, directly or metaphorically."

When color speaks to us, how well do we listen? How do we respond? Do we *reflect upon* and *reflect* its truth? Or do we bend and conceal it? The enormous power of art resides in its ability to reveal truth. In its ability to sensitize the eyes of our heart, that we may be better able—to paraphrase Ralph Waldo Emerson—to detect that gleam of light that flashes from within.

CHAPTER FOUR

RHYTHM OF THE SUN

The earth cannot move without music. The earth moves in a certain rhythm, a certain sound, a certain note. When the music stops the earth will stop and everything upon it will die.
— *Sun Ra* —

Everything in the universe has a rhythm, everything dances.
— *Maya Angelou* —

I t was as if they could hear its music. It was pulsating 93 million miles away, and its enormous sound could not permeate the vacuum of space, yet they danced as though they could hear and feel its complex rhythms. Dozens of them gathered in the remote clearing. They were exhausted, in pain, uncertain about the future. Beyond the clearing, a living nightmare awaited. Yet they believed there was reason to hope.

They danced around the circle counterclockwise, mirroring the Sun's daily east-to-west motion across the sky, as seen in the Southern Hemisphere.

The rising of the Sun represented birth; its arc across the sky, life; its setting, death. Water symbolized the afterlife, and a tree or a cross stood for the vital connection between this life and the next, the realm of the ancestors. The ritual incorporated singing, percussion, and dance; as the historian Sterling Stuckey explains, in his seminal book *Slave Culture*, dance was "primarily devotional, like a prayer."* This time, the ritual was prompted by a recent death, and just outside the circle, under a tree, was a burial mound decorated with the beloved's belongings and broken pieces of glass. As dusk fell, the dancers' steps quickened to match the pace of the drumming; their chanting reverberated with increased intensity into the sky, until they became completely transported and could at last transcend their harsh reality.

> I walk in de moonlight, I walk in de starlight,
>> To lay dis body down.
> I'll walk in de graveyard, I'll walk through de graveyard,
>> To lay dis body down.
> I'll lie in de grave and stretch out my arms;
>> Lay dis body down.
> I go to de judgement in de evenin' of de day,
>> When I lay dis body down;
> And my soul and your soul will meet in de day
>> When I lay dis body down.

One of the most widespread and enduring rituals performed by enslaved Africans in North America, the *ring shout* was a dance consecrating the cycle of birth, life, and death. In response to a birth or death, a community would gather in secret to celebrate the passing of a loved one to a better world or to mourn a new soul's entry into this one. The Sun stood at the center of this sacred ceremony, which synthesized elements of the various traditions the Africans had carried across the Atlantic.

---

* Leonardo Boff echoes this, when he says in *Cry of the Earth, Cry of the Poor*, "In Asian and African traditions dance is the liveliest and most concrete symbolic representation of the world."

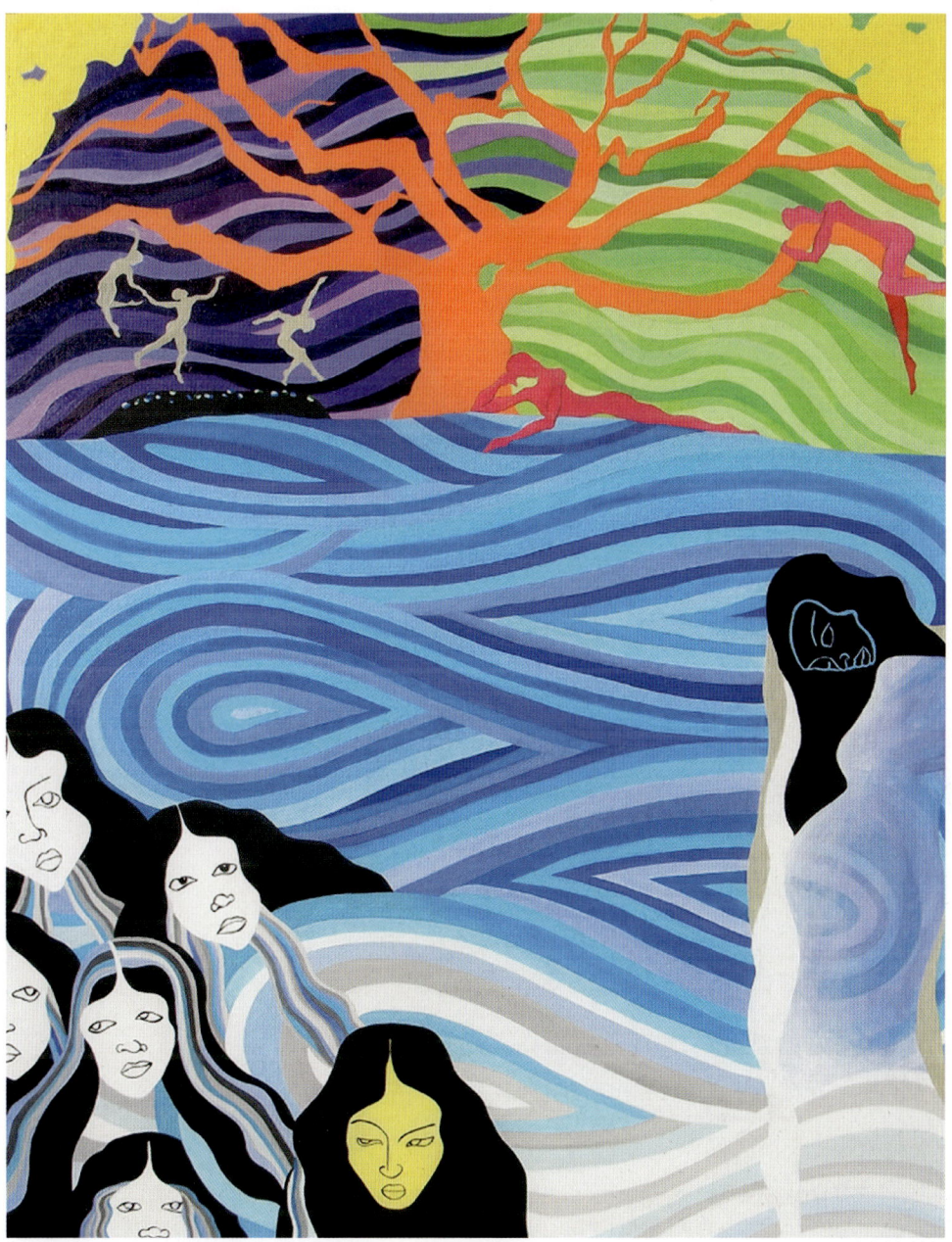

SHOUT | Nia Imara, 2013, oil on canvas, 28 × 22 in (71.1 × 55.9 cm). I created this piece as an homage to the ring shout.

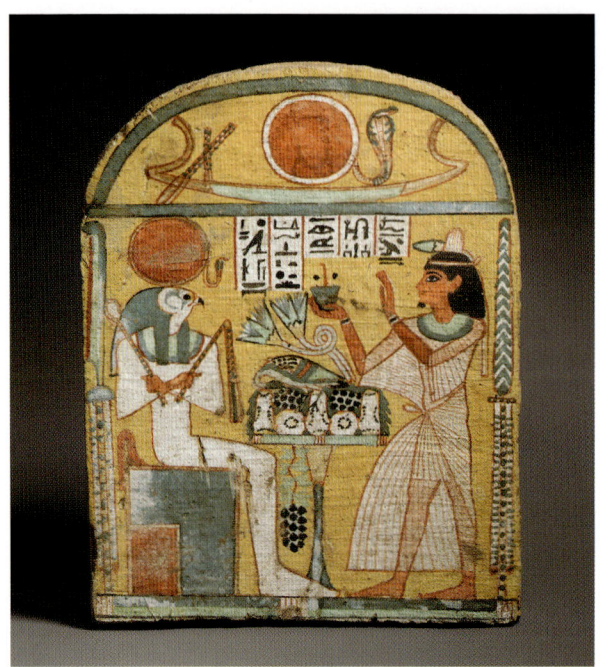

STELA OF AAFENMUT | Egypt, ca. 926-889 BCE, wood, gesso, paint, 9 1/16 × 7 3/16 × 1 3/8 in (23 × 18.2 × 3.5 cm). Aafenmut (standing right) offers incense to the Egyptian Sun god, Re-Harakhty (seated left).

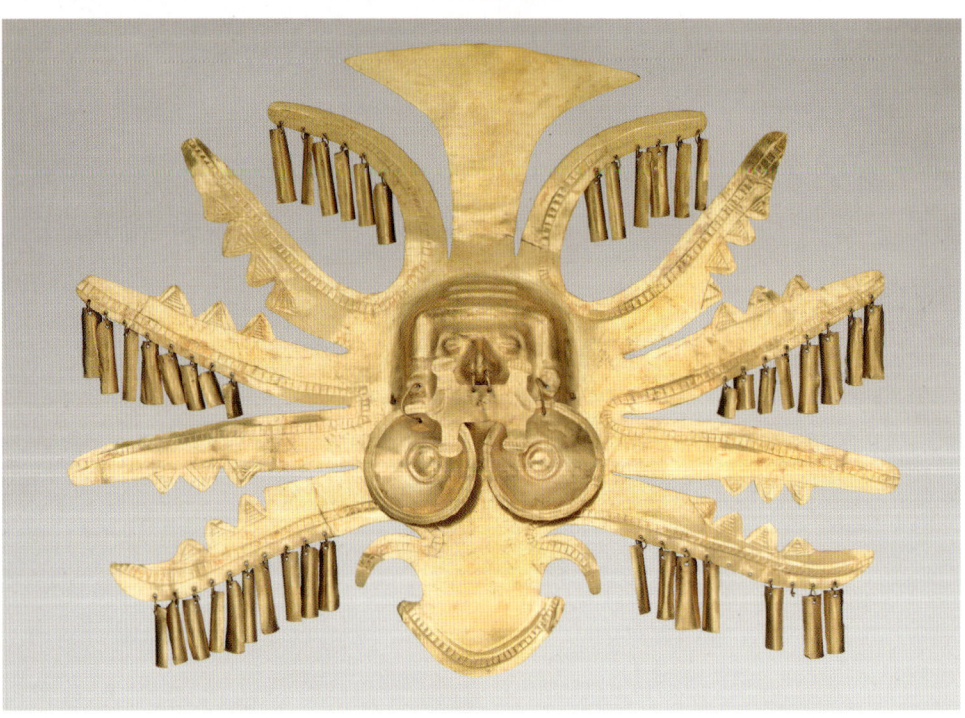

GOLD HEADDRESS ORNAMENT | Calima (Yotoco), first–seventh century, 8 1/2 × 11 1/2 × 1 1/4 in (21.6 × 29.2 × 3.2 cm). In Calima cultures, gold was symbolic of the Sun's creative power.

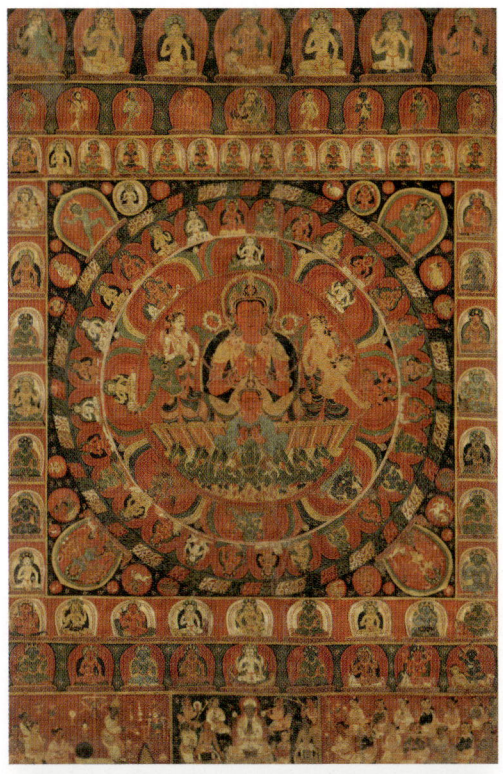

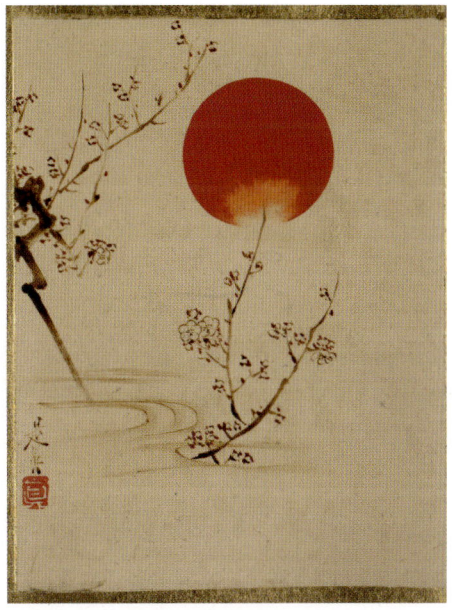

SUN AND PLUM BRANCHES | *Top right:*
Shibata Zeshin, Japan, Edo period (1615–
1868), album leaf; lacquer on paper, 4½ ×
3½ in (11.4 × 8.9 cm).

MANDALA OF THE SUN GOD SURYA | *Top
Left:* Kitaharasa, Nepal, likely 1379,
distemper on cotton, 32⅝ × 21½ in (82.9 ×
54.6 cm).

**CORONATION STONE OF MOTECUHZOMA II
(STONE OF THE FIVE SUNS)** | *Bottom Left:*
Aztec (Mexica), 1503, basalt, 22 × 26 ×
9 in (55.9 × 66 × 22.9 cm). Art Institute of
Chicago.

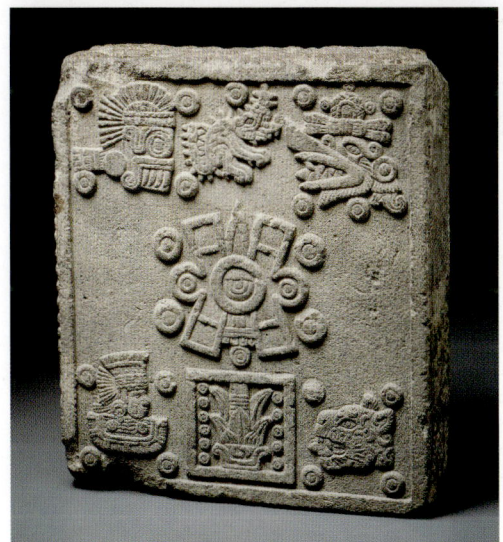

The ring shout echoes through the soul of America's most original art forms—blues and jazz. Like the light of the Sun vibrating through space, the rhythms of the ring shout can be felt rippling through John Coltrane's "changes" and Alvin Ailey's *Revelations.**

The creators of the ring shout shared the deep reverence for the Sun common to many peoples. Our Solar System's star is venerated in visual art, music, poetry, dance, architecture, and often as a manifestation of religious belief. From the Mesopotamians to the Aztecs, people in different cultures have related to the Sun as a symbol of power, goodness, divinity, and life.

The Dynamic Sun

One reason why we're so strongly drawn to the Sun is because we're stuck to rhythm like white on rice. From our circadian rhythm, our sleep-wake cycle over the course of 24 hours; to the carbon cycle, which keeps our planet pleasantly warm; to the daily rising and setting of the Sun; to our most important internal rhythm, our heartbeat, life dances to a rich tapestry of pulses that are so steady and consistent, we forget they're there. Rhythm is inherent to life, and the lead drummer for life on Earth is the Sun.

Our closest star is itself a paragon of rhythm. *Many* rhythms, in fact. As the planets orbit the Sun, the Sun orbits the center of the Milky Way, dragging the entire Solar System along with it. One revolution lasts roughly 250 million years, so we've made almost 20 laps since the birth of the Solar System 4.6 billion years ago—it's a slow but steady dance.

Also like the planets, the Sun turns on its axis. But because it's a gigantic ball of gas, it doesn't rotate rigidly like the rocky planets or the Moon.[†] It actually spins faster at the equator (rotating once about every 25 days) than at the

---

[*] "Changes" refer to Coltrane's unique treatment of harmonic progressions, inspired by the work of music theorist Nicolas Slonimsky.

[†] The rocky planets—Mercury, Venus, Earth, and Mars—are the closest planets to the Sun and made mostly of rocks and metals. The gas giants—Jupiter, Saturn, Uranus, and Neptune—are composed mostly of hydrogen and helium.

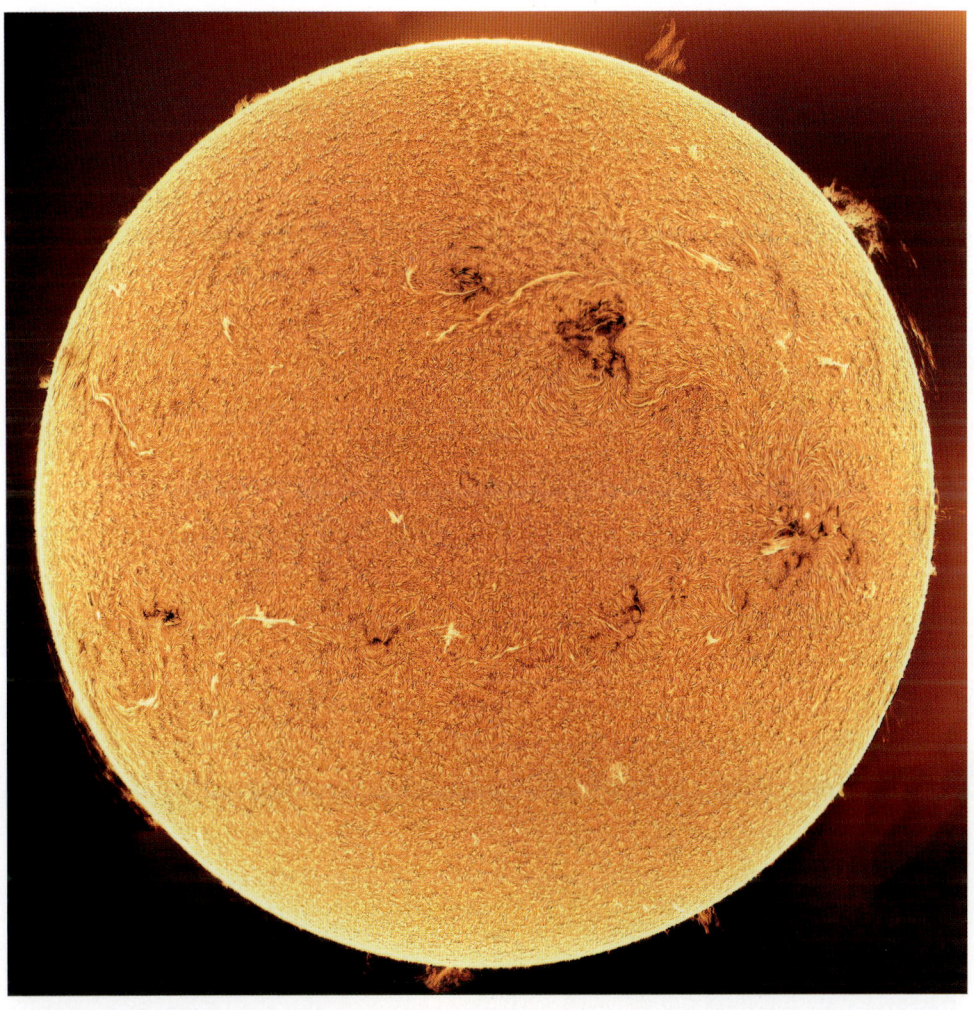

THE ACTIVE SUN | Sunspots and prominences—loop-like structures rising from the surface of the Sun—freckle the face of our most familiar star.

poles (once every 38 days). Scientists measure the rate of spin by tracking long-lived sunspots, the dark blotches, typically appearing in pairs, sprinkled across the surface of the Sun. Sunspots can be as large as or bigger than Earth and can last for hours, days, or weeks. Why do they appear dark? The Sun is constantly churning, with enormous pockets of gas relentlessly bubbling up from below the surface, like a gigantic pot of oatmeal on the stove. Strong magnetic fields curling throughout the star occasionally suppress the rising of gas, effectively jamming the outward flow of energy and creating sunspots. Since there is less energy (that is, less light) in these areas, sunspots are cooler, by at least a couple of thousand degrees, than their surroundings. Therefore they appear darker.

In 1825, a 36-year-old German pharmacist-turned-astronomer named Samuel Schwabe won a telescope in a lottery. A few months later, he purchased a bigger telescope and built an observatory from which he began searching for planets inside the orbit of Mercury. Though he never found one, he carefully monitored and recorded sunspots. Over the years, Schwabe noticed how their positions and numbers changed, and in 1843 he announced a discovery: a 10-year cycle in the appearance of sunspots. Other astronomers didn't take him seriously right away, but it turns out that Schwabe was pretty much *spot* on. Now that we have records going back more than 400 years, since Galileo first pointed his telescope at the Sun (not knowing it could harm his eyes!)[*] and documented sunspots, the evidence is now clear that it's actually an *11*-year solar cycle.

At the peak of a cycle, when sunspots are more numerous and intense, the surface of the Sun ripples with activity, such as powerful explosions, storms, and eruptions of scorching gas. During low points, the number of sunspots plummets; they may even vanish altogether. Astronomers don't yet fully understand the origins of the solar cycle, which can vary unpredictably in its intensity, but they do know that it's driven by the Sun's magnetic fields. The entire Sun, in fact, is a gigantic magnet. A very complicated, incredibly

---

[*] Fortunately, Galileo did not injure his eyes by observing the sun. More than 20 years after these observations, he did go blind, though the likely cause was cataract and glaucoma.

dynamic magnet. Roughly every 11 years, when solar activity is at its maximum, the north and south poles of the Sun's magnetic field switch sides, like a celestial gymnast executing a cosmic half somersault!

When the Sun is very active, we can feel significant impacts on Earth—though there is no established connection between solar activity and long-term climate change. A Sun-driven surge in "space weather" can, however, temporarily disrupt satellites and even endanger astronauts. Here's a rather dramatic example. In 1967, when a tremendous solar storm jammed some communication signals, the United States nearly started a war with the Soviet Union. Having prematurely concluded that Russian interference had caused the problem, the government put the US Air Force on alert. At nearly the last moment, it was revealed that the real culprit causing the signal interference was space weather; thus World War III was narrowly prevented. Probably not what Stevie Wonder had in mind when, a few years later, he recorded "Blame It on the Sun"!

"Who Has Prepared It All?"

The consistency of the Sun's rhythms is the basis for life, yet it is full of surprises that keep us wondering. In music, too, we are drawn to steady rhythms, but it is the occasional breaks in pattern that keep us engaged. Musical geniuses like James Brown, J. S. Bach, Louis Armstrong, Ella Fitzgerald, Thelonious Monk, and Stevie Wonder have always known this. Bach's Two-Part Inventions and Stevie Wonder's "Superstition" alike mesmerize us with their hypnotic beats, but it is the nuanced variations in the overall patterns that really capture our interest. The same is true in visual art—pattern and repetition are the visual equivalent of rhythm.

The vivid, living colors in the paintings of Alma Woodsey Thomas absorb us for some of the same reasons why "Superstition" is so irresistible. In a piece such as *The Eclipse*—from her series of *Space Paintings*—Thomas's use of color, brushstroke, shape, and space establish her visual tempo. Coherence is achieved with the overall repetition of the circular patterns—the visual "beat"—while the palettes are unified because they are composed mostly of primary colors of all the same value (meaning their level of lightness or

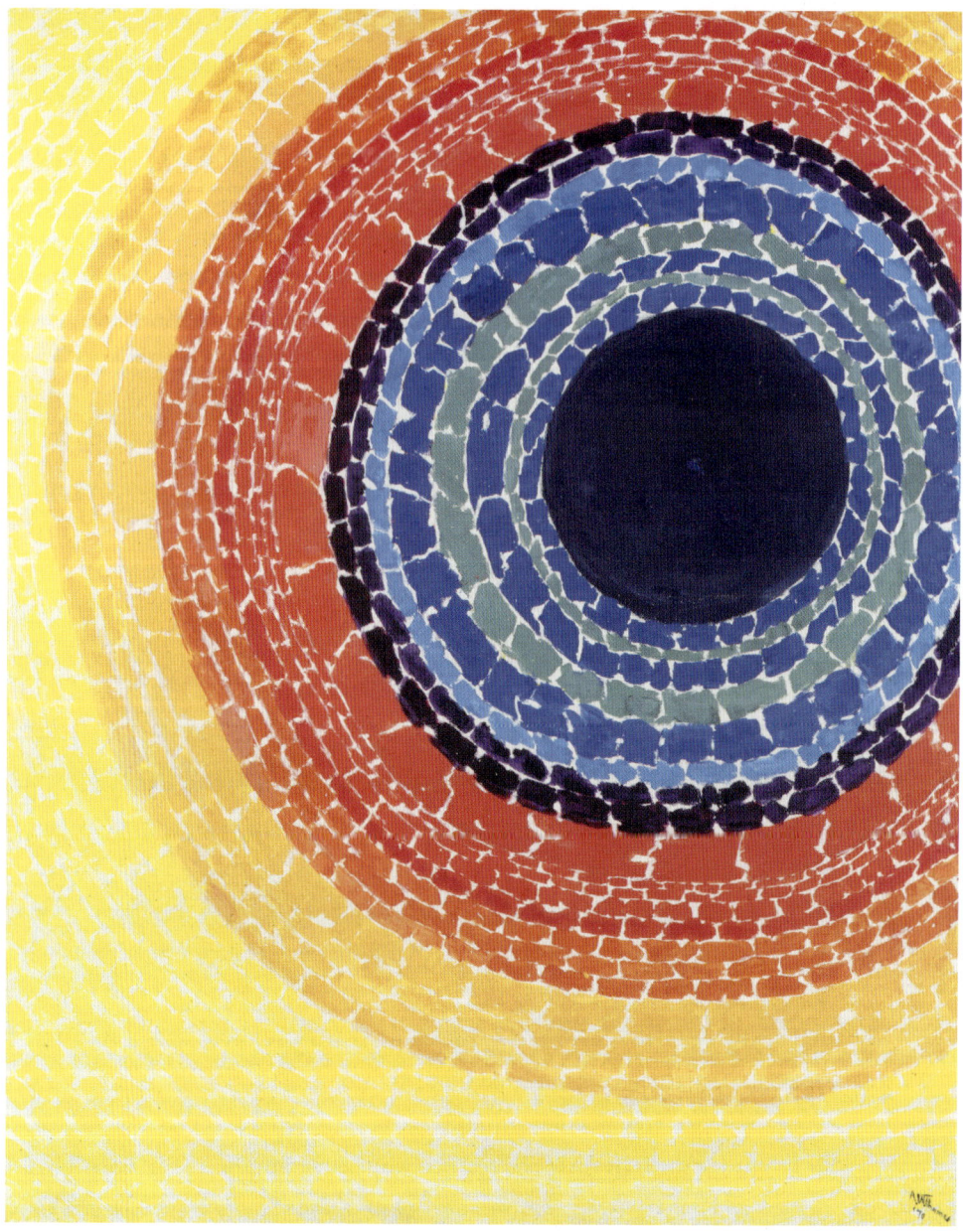

THE ECLIPSE | Alma Thomas, 1970, acrylic on canvas, 62 × 49¾ in (157.5 × 126.4 cm). Thomas was inspired by the total solar eclipse of March 7, 1970, which was visible from the East Coast of the US. A solar eclipse occurs when the Moon passes between the Sun and Earth and blocks the Sun's light—partially or fully—from some parts of Earth.

darkness). In *The Eclipse*, the blue and red rings radiating from the center pull the eye as if toward a bull's-eye; but they are not all the same blues and reds. I count at least four or five different shades of each hue. The individual blocks of color composing each ring are roughly the same in size and hue, but there are slight variations. The thickness of the rings also changes. Particularly arresting is the thin crimson one, which boldly leaps out from the middle of the red group, creating a sense of movement. Just like a Stevie Wonder song or a work by Bach—or the Sun itself—there are *multiple* rhythms dynamically interacting across the canvas.

The early paintings of the Georgia-born teacher and artist are in the style of realism, but the later works she became famous for are distinguished by elements of abstract expressionism, the artistic movement that Mark Rothko and Wassily Kandinsky, her contemporaries, are associated with. Rothko loved music and found in composers like Mozart and Haydn artistic principles that he applied to his deceptively simple paintings. Kandinsky, for his part, saw painting as like creating music and achieved rhythm in his own inimitable way through the unpredictable repetition of forms.

Thomas's work, too, stands out for its musicality. She was inspired by the music of *nature*, as we can see by the titles of many of her works, pieces with names like *Étude in Brown*, *Red Roses Sonata*, and *White Daisies Rhapsody*. She once commented on how she achieved the rhythmic quality of her work, saying, "My strokes are free and irregular, some close together, others far apart, thus creating interesting patterns of canvas peeking around the strokes." I can imagine how the process of painting was as trance inducing as dancing for Thomas, who in 1972, at the age of 81, became the first black woman to hold a solo exhibit at the Whitney Museum.

Going back much further than the advent of abstract expressionism are the artistic traditions of the Bwa people of West Africa. Though separated from their relations in North America by the Atlantic, there is no doubt that many cultural bonds survived the devastation of the Middle Passage. The Bwa's stunning plank masks, known as *nwantantay*, have a striking resonance with the paintings of Alma Thomas. The Bwa use these masks in rituals related to agriculture, childbirth, initiation, and burial. The large, intricately

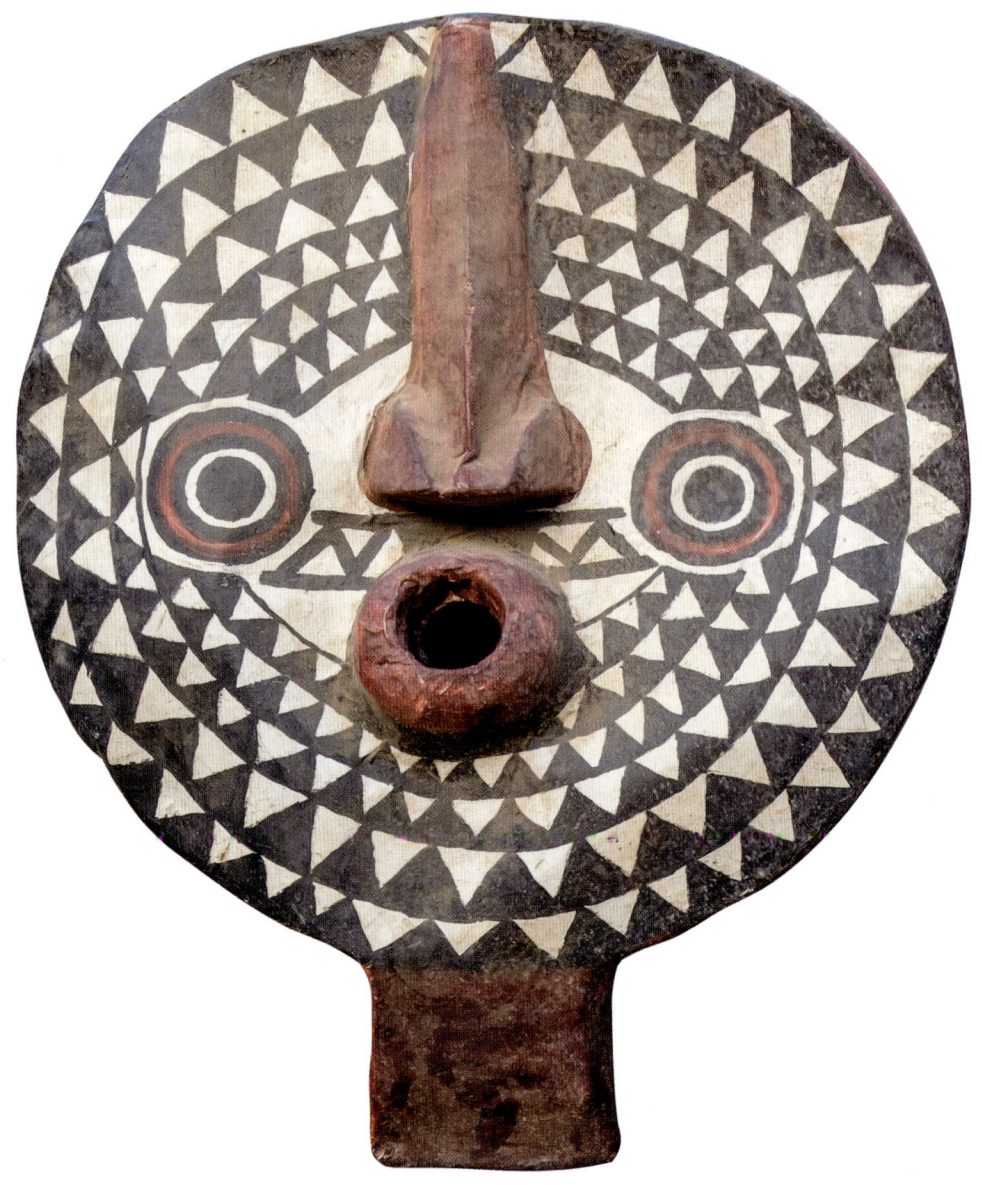

NWANTANTAY | A Bwa mask representing the Sun. *Credit: Hemis / Alamy Stock Photo.*

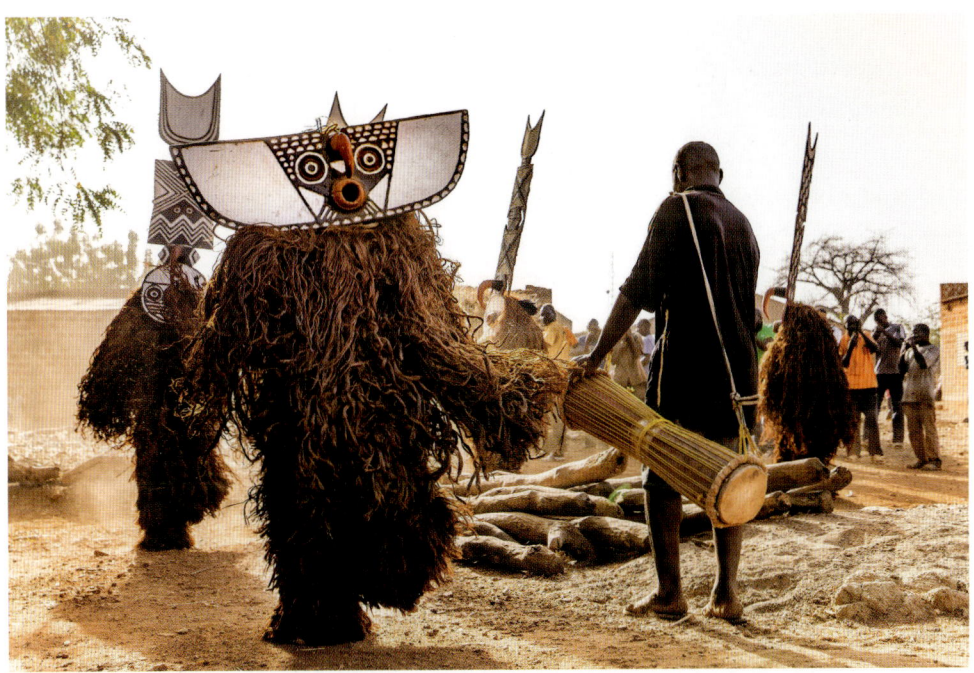

A BWA CEREMONY | The dancers wear abstract masks representing the spirits that animate them.

Credit: Elena Bobrova / Alamy Stock Photo.

carved wooden masks take many forms—the Sun, animals, insects—each symbolizing the spirit that animates the particular mask. Like much African art, the masks are highly abstract. *Nwantantay* are painted with crosses, zigzags, and checkerboards in white, black, and red. White often symbolizes ignorance or death, with black representing knowledge and purity, and red, life. Thus, the black-and-white checkerboard and cross patterns symbolize the intermingling of youth and age, knowledge and ignorance.

Art is a communal affair in a Bwa village. Everyone—young and old, male and female—has access to the masks, though only the men wear them during rituals. The masks, often several feet tall, are worn with elaborate, vibrantly colored costumes made from the long fibers of the raffia palm tree. In preparation for initiation ceremonies, the young male initiates work together to restore the costumes and masks. Under the guidance of an elder, they meticulously retouch any faded paint on the masks with natural pigments, careful to preserve the integrity of the design. Once the spirits have been acknowledged and received sacrifices, the ceremony can begin. The masks are donned, women gather in groups according to age, the musicians take up their instruments. The ritual can last for hours. With the visual echoes of the black-and-white *nwantantay*, the drumming, the patterns on the women's dresses, the dancing in circles, it is an explosion of rhythm.

Now let's leap across a continent to 19th-century Russia, where the great writer Leo Tolstoy venerated the Sun and nature, too. Celebrated for his epic novels, Tolstoy also wrote a great deal of pedagogy. He was especially concerned with the education of the peasantry. He considered his school primer, *Azbuka* (Russian for *ABCs*), the most important work of his life, more so than his wildly popular sagas *War and Peace* and *Anna Karenina*. He also wrote many short pieces about science and the natural world for children, including "The Sun's Heat," a tender portrait of the Sun and our dependence on it in nearly every aspect of our lives.

> The grass, the grain, the fruits, the trees grow up; animals find their food, men eat their fill, and gather food and fuel for the winter; they build themselves houses, railways, cities. Who has prepared it all?— The Sun.
>
> A man has built himself a house. What has he made it of? Of timbers. The timbers were cut out of trees, but the trees are made to grow by the Sun.
>
> The stove is heated with wood. Who has made the wood to grow?—The Sun.
>
> Man eats bread, or potatoes. Who has made them grow?— The Sun. Man eats meat. Who has made the animals, the birds to grow?—The grass. But the grass is made to grow by the Sun.
>
> A man builds himself a house from brick and lime. The bricks and the lime are burnt by wood. The wood has been prepared by the Sun.
>
> Everything that men need, that is for their use,—all that is prepared by the sun, and on all that goes much Sun's heat. The reason

that men need bread is because the Sun has produced it, and because there is much Sun's heat in it. Bread warms him who eats it.

. . . Watermills and windmills turn around and grind. Who moves them?—Wind and water. And who drives the wind?—Heat. And who drives the water?—Again heat. Heat raises the water in the shape of vapor, and without this the water would not be falling down. A machine works,—it is moved by steam. And who makes steam?—Wood. And in the wood is the Sun's heat.

Heat makes motion, and motion makes heat. And both heat and motion are from the Sun.

Much has changed since Tolstoy's time—and much has not. Steam power still generates most of the world's electricity. These generators are no longer fueled by wood but rather by coal, natural gas, or oil, which are ancient reserves of the Sun's collected energy, stored in plants and animals that died long ago.

Tolstoy's deceptively simple story is infused with a sophisticated sense of rhythm. He mirrors the cadences of the Sun in the style of the writing itself, thus echoing and amplifying the overall rhythm. This, combined with the way he personifies and personalizes the Sun, produces a feeling of tenderness for our star. Art can make us feel more deeply connected to nature. The same elements are used in a Bwa ritual—layered rhythm, personification, personalization. The joint effect is to create a sense of intimacy with our powerful magnetized star.

Aura of the Sun

The Sun has amazing power. An average star suspended in space on the outskirts of the Milky Way Galaxy, it supplies us with an overflow of energy, generates weather patterns, drives the ocean currents, and feeds the thousands of species of plants (nature's light eaters) that cover the planet and provide us with oxygen and food.

What exactly *is* that burning light in the sky?

The Sun is an *enormous* ball of *hot, ionized gas, shining* due to its own *power*. Let's break down each of these terms, starting with the last.

Intuitively, we recognize the Sun's incredible power, which scientists define as the rate at which energy is produced or used. The Sun releases prodigious amounts of energy in the form of light. Every second, in fact, it sheds enough light to provide for the world's annual energy consumption for 700,000 years!

The Sun generates so much energy because it's so darn big. This behemoth at the center of the Solar System contains more than 99 percent of the system's mass and anchors the planets with its gravity. In the 3rd century BCE, the Libyan polymath Eratosthenes estimated the size of our star as 27 times bigger than Earth. He was a little off. Though many of his other calculations—for example, concerning the size of Earth—were correct, today we know that the Sun is *109 times bigger* than our planet. In other words, 109 Earths could line up across the diameter of the Sun. Looked at another way, the Sun is so gigantic that 1.3 million Earths could fit inside it!

The Sun's immense gravity produces extreme pressure at its center, which creates the conditions necessary for nuclear fusion, the source of its power. In nuclear fusion, atoms combine and release photons—those particles of light—in the process. Fusion is what causes our Sun and most other stars to shine steadily for billions of years.

Like all stars, the Sun is made mostly of hot hydrogen gas, an idea first proposed in 1925 by Cecilia Payne-Gaposchkin in her pioneering PhD thesis. Thousands of miles into the Sun's interior, at its heart, the temperature is a blistering 27 million degrees Fahrenheit. This is more than hot enough to strip the electrons off atoms, thus forming a hot, roiling, electrically charged gas called plasma. Hydrogen atoms are constantly smashing into each other and fusing, thus making helium atoms (four hydrogen atoms are needed to make one helium atom). A helium atom is, counterintuitively perhaps, less massive than the four hydrogen atoms it took to create it. That extra bit of mass gets converted into light during the fusion process.

The Sun is crammed with hydrogen, and the photons generated at its center repeatedly collide with other particles as they try to fly to the surface,

CROSS SECTION of the SUN

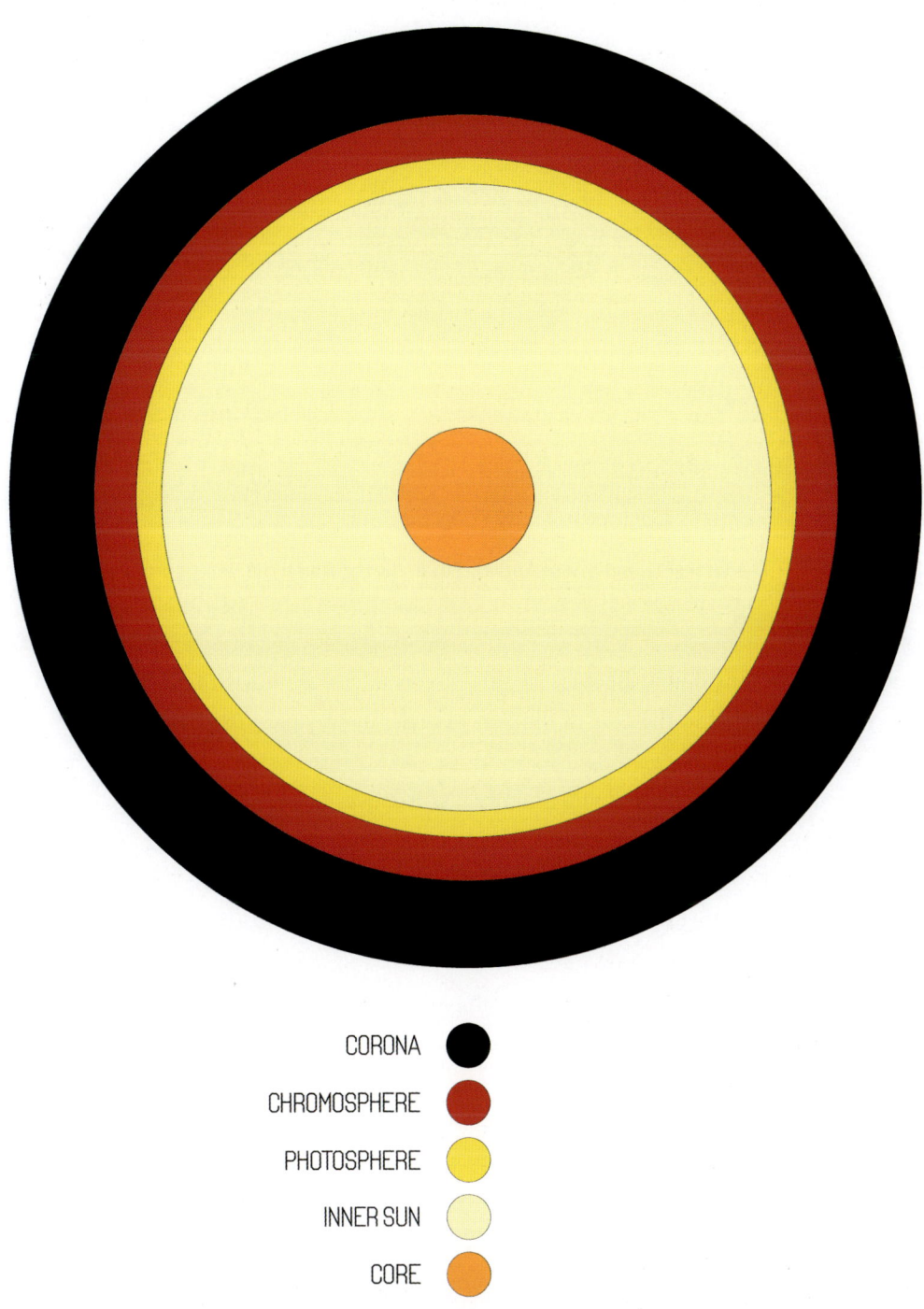

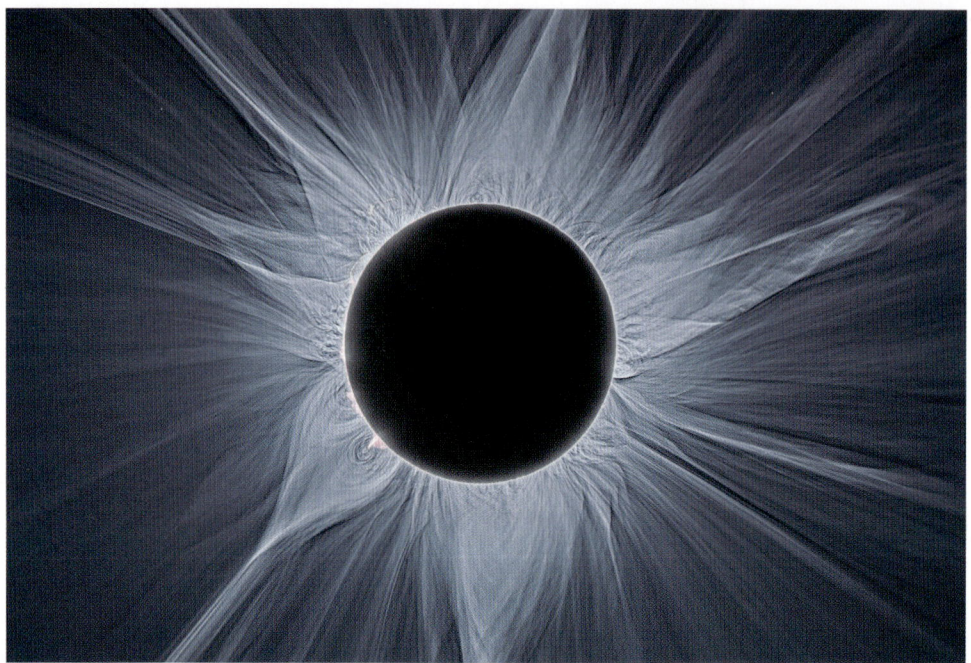

TOTAL ECLIPSE | The total solar eclipse of April 8, 2024, during which the Sun's corona was visible.

slowing the photons down. Because of this, it can take *more than 100,000 years* for light created in the core of the Sun to make its way out. *That's* how big the Sun is.

Let's follow one of these flying photons. As it moves outward from the center, the temperature drops, until it reaches the photosphere—meaning "sphere of light"—the surface of the Sun. The photosphere is where the majority of the light that's produced in the interior escapes. It's the part of the Sun we can see. It is the thinnest and lowest layer of the Sun's atmosphere, about 300 miles (500 kilometers) thick. Compared to the center of the Sun, it's a mild 10,000 degrees Fahrenheit, which is still 20 times hotter than an oven turned to its maximum setting!

Imagine warming yourself next to a campfire. As you walk away, something strange happens—you feel the air around you getting *hotter*. The Sun is like this magical campfire. Surprisingly, outward from the photosphere, the temperature *increases*.

Next comes the chromosphere, a thin layer only about a thousand miles thick and rising in temperature to some 14,000 degrees. The gas in the chromosphere is very sparse; its distinctive reddish color is visible from Earth only during a total solar eclipse. While observing the total eclipse of 1868 in southern India, the French astronomer Jules Janssen identified a peculiar new element in the chromosphere. It clearly wasn't hydrogen but a different atom that hadn't before been detected on Earth. A couple months later, British scientist Norman Lockyer also observed helium in the Sun and named the element after the Greek word for Sun, *helios*. Helium, that hilarious atom that makes balloons fly and our voices squeak, is the second most abundant element in the Sun.

Beyond the chromosphere, the temperature suddenly spikes to over a million degrees. Welcome to the corona, the outermost and hottest layer of the solar atmosphere. Since the gas in the corona is sparse—more than a million times less dense than the photosphere—it is very faint. In fact, the Moon is about twice as bright as the corona, whereas the photosphere outshines the Moon most of the time. Like the chromosphere, the corona can be seen only during a total solar eclipse—and during those brief, dreamlike minutes, the weblike structure of the corona, shaped by the Sun's magnetic fields, is revealed.

Why the Sun's corona is thousands of times hotter than its surface remains a mystery. Many astronomers believe it has something to do with how the complex magnetic fields blanketing the Sun release energy. We do know that the corona extends millions of miles into space and is the source of solar wind, a high-speed torrent of charged particles streaming through interplanetary space. The corona and solar wind together fill the Solar System. We're all, quite literally, living in the aura of the Sun.

Sun Song

The farthest reaches of the Solar System are touched by the Sun's silent light. But the Sun itself is anything but quiet. In fact, it's incredibly noisy.

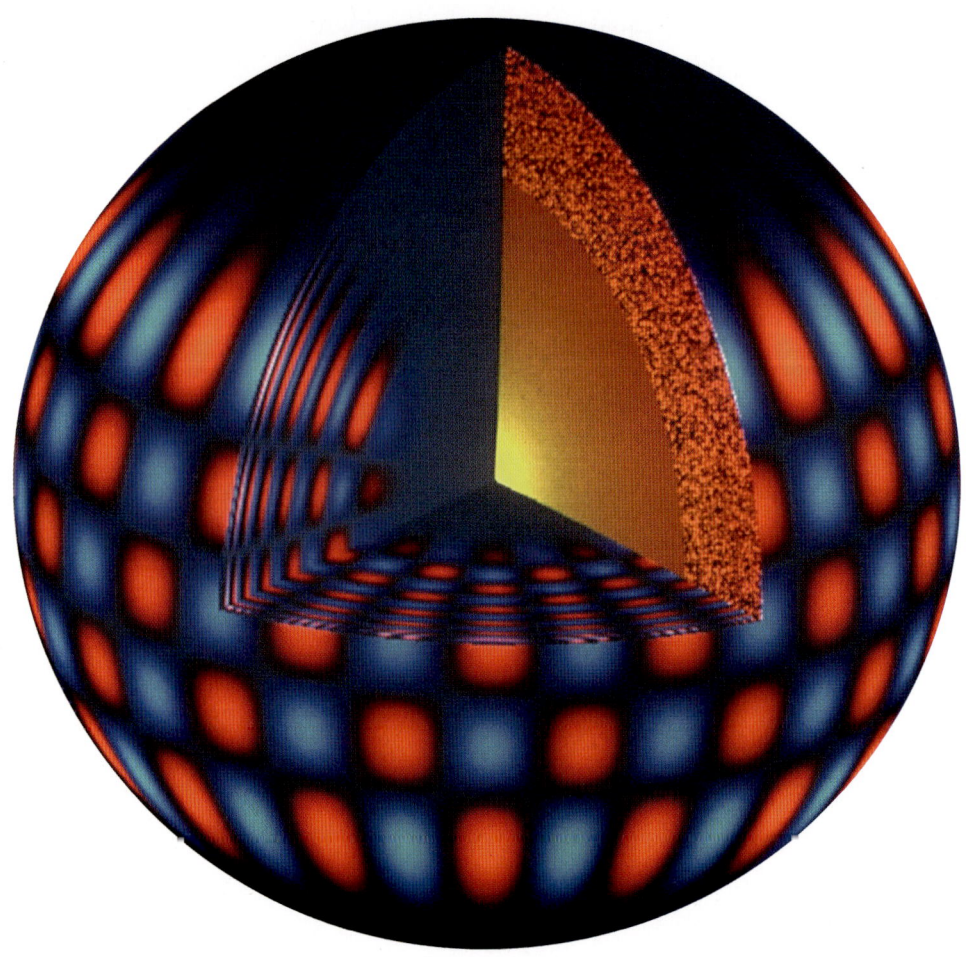

SOUNDS OF THE SUN | A computer-generated image of one of the millions of modes of sound wave vibrations of the Sun. Regions that are approaching and receding from the surface are shown in blue and red, respectively. *Credit: NSO/AURA/NSF.*

Though it may seem strange, the surface of the Sun is virtually opaque to light. Its brilliant halo hides much of what lurks below the surface. Although the vast ocean of plasma beneath the photosphere is all but invisible, there is much about it that we can understand. What we cannot see, we can *hear*. We can *listen* to its heartbeat. The Sun pulses with sound.

The interior of our star bubbles and churns with hot, massive pockets of gas that continuously surge to the surface at thousands of miles per hour, where they eventually break and spew their stored energy. The cooled gas rains back down, heats up, and rises again, in a continuous cycle. The relentless crashing of material against the surface makes the Sun vibrate like a celestial drum.

Yet the Sun's song is effectively blocked at the surface. Compared to air on Earth, interstellar space is sparse, and unlike light, sound waves need a medium through which to travel. Scientists have estimated that if interstellar space were filled with air, the Sun's song would reach Earth at a volume of about 100 decibels, the equivalent of a honking car horn or hair dryer positioned right next to you. And that's how it would sound from 93 million miles away! Just imagine—if we could stand near its surface, the Sun would be deafening.

Helioseismology is the science of listening to the Sun. Analogous to how geoseismologists learn about the interior of our planet by "tuning in" to earthquakes, solar physicists study the Sun by tuning in to *sunquakes*, vibrations that carry information about the Sun's dynamic interior. With its multitude of complex rhythms and overlapping notes, the Sun rings out like a piano with millions of keys. Helioseismology has been likened to trying to figure out how a piano is constructed based on the sounds it makes while plummeting down a stairway.

Mercifully, scientists haven't yet invented the microphone or recorder that we can place right next to the Sun to hear its lurid song. Rather than actually "listening" to the Sun, scientists observe the changes in light that appear on the surface as a result of the constant ebb and flow of material from below. This is the Doppler effect, the change in frequency of waves produced by moving objects. In this case, the Sun's vibrating surface produces changes in the frequencies of light waves emanating from it.

In addition to the surprising loudness of the Sun's song, its voice has a distinctive timbre. Scientists have found that the Sun croons in a deep, *deep* baritone. Its voice falls in a frequency range of about 5 million hertz, far below the range the human ear is sensitive to. By translating the Sun's tones to pitches the human ear can hear, scientists have produced recordings of the Sun's song. Imagine a low, prolonged rumble, or the dull roar of the ocean. The Sun sings in an ancient language not entirely intelligible to human ears, but it can stir the soul and the imagination just the same.

Sometimes the artist intuits truths about nature as profound as the discoveries of the scientist. In his 1927 poem "Sun Song," Langston Hughes's repetition of the word "sun" echoes the star's own rhythms, much like a painting by Alma Thomas or a story by Leo Tolstoy.

> Sun and softness,
> Sun and the beaten hardness of the earth,
> Sun and the song of all the sun-stars
> Gathered together—
> Dark ones of Africa,
> I bring you my songs
> To sing on the Georgia roads.

Decades before the Sun's oscillations were first detected in the 1960s, Hughes perceives that all of space silently vibrates with the songs of stars. He gives voice to the Sun's quiet beckoning, to its invitation for the "Dark ones of Africa" to join it in song.

All art is essentially metaphor. Through my experience as both scientist and artist, I've discovered that one of the most beautiful and powerful things nature has to offer us is its metaphors. It sometimes amazes me to recall that everything I see is a reflection of the Sun. It fills me with wonder that

I, too, am a reflection of the Sun. Even more astonishing, I'm its walking embodiment—one among eight billion—because everything that I've ever put inside, on, or around my body drew its energy, directly or indirectly, from the Sun.

Then I'm reminded that, just as everything is a reflection of the Sun, everything is a reflection of *me*, of my perceptions. My beliefs, emotions, and attitudes are the mirror upon which reality is reflected. For better or worse, I see the world by the light with which I illuminate it. As Ralph Waldo Emerson said, "Each of us sees in others what we carry in our own hearts."

A government bent on war and defense will see a potential enemy in everything, including phenomena caused by the innocent Sun. An enslaved people see in the same Sun a powerful symbol that connects them with home, with their ancestors. It offers a perpetual reminder that life occurs in cycles—rhythms—and thus, their peculiar situation is not permanent.

The rhythm of life. It's a potent metaphor. Art has the unique capacity to draw from below the surface rhythms inherent within us and in nature. When we paint, write, sing, or dance with the Sun, we intensify these rhythms. We love rhythm because it brings us into sync with ourselves, with one another, with life itself.

CHAPTER FIVE

FINDING LIFE

The most beautiful experience we can have is the mysterious. It is the fundamental emotion which stands at the cradle of true art and true science. Whoever does not know it can no longer wonder, no longer marvel, is as good as dead, and his eyes are dimmed.
— Albert Einstein —

Art evokes the mystery without which the world would not exist.
— René Magritte —

For the secret of man's being is not only to live but to have something to live for.
— Fyodor Dostoevsky —

In 2022, a team of researchers announced that they had completed a full mapping of the human genome. It was the culmination of an effort begun over three decades earlier with the Human Genome Project. Scientists could now claim to describe the precise arrangement of the thousands of genes composing the blueprint for human life, DNA. Each of the trillions of cells in our body carries this personal instruction manual. Humans share 96 percent of it with gorillas, 60 percent with fruit flies, and half with trees. DNA contains all the directions needed for living organisms to function, grow, and reproduce.

Yet a map of the human genome is rather like a map of galaxies in the universe. Just as DNA is found in all life, stars are found in all galaxies ever observed. We know what stars are and what they do, but what is a galaxy? Galaxies are made mostly of *dark matter*, an unknown substance that pervades the cosmos, yet its nature is a complete mystery to astronomers. Similarly, we know what DNA is and what it does to enable life, but *what is life itself*?

How did life start? When does life begin? These seem like basic questions, but our answers are anything but. They are the opening lines to one of the greatest mysteries.

Biologists have debated the definition of life for a long time but still don't agree. Why have they had such a hard time? In large part this is due to the incredible diversity of life, which resists being arranged neatly in a box. There is, however, general agreement about some of its characteristics. Most life as we know it consists of cells and is capable of growth, reproduction, maintaining stability (homeostasis), converting food into energy (metabolism), and adaptation. Yet as scientists consider borderline cases— like viruses, sterile animals, and artificial intelligence—a definition of life remains elusive.

Some argue that maybe we shouldn't bother trying to come up with a single comprehensive definition. The poets and philosophers throughout the ages who have tried to capture life's mysterious essence with metaphor might agree. Sai Baba of Shirdi said, "Life is a song—sing it. Life is a game—play it. Life is a challenge—meet it. Life is a dream—realize it. Life is a sacrifice— offer it. Life is love—enjoy it."

Langston Hughes, meanwhile, reminds us in "Dreams" that life's needs are more than material:

Hold fast to dreams
For if dreams die

Life is a broken-winged bird
That cannot fly.

Hold fast to dreams
For when dreams go
Life is a barren field
Frozen with snow.

Scientists have held fast to the dream of mapping the human genome and the cosmos itself. Using data from the Sloan Digital Sky Survey, an ambitious program to map the universe, astronomers have pinpointed the locations of hundreds of thousands of galaxies, each containing billions of stars and planets. Do any of those planets harbor life? Or is the DNA-based life that we're familiar with here on Earth unique in the universe? In other words, are we alone?

This question drives the search for *exoplanets*—planets beyond our Solar System. Up until a couple of generations ago, we could only speculate about the existence of other worlds within our Galaxy. But since the first exoplanet was discovered, in 1992, thousands have been confirmed. In fact, astronomers now have good evidence that billions of Earthlike planets circle Sun-like stars here in the Milky Way. But if life flourishes elsewhere in the cosmos, it may exist in a form and place that neither scientists nor science fiction writers have yet imagined. Given the infinite possibilities, our best shot at finding it is to start by looking for *life as we know it*.

Hot Jupiters, Mini-Neptunes, and Super-Earths

The story of exoplanet discovery is essential background to understanding the current search for life beyond Earth. And it begins with some surprises. In 1992, the astronomers Aleksander Wolszczan and Dale Frail serendipitously found the first planets outside the Solar System. They weren't where you might expect. Instead of revolving around a "normal" star like our Sun, the planets were found to be circling a pulsar—a rapidly spinning, exceedingly dense relic of a massive star that exploded as a supernova.

The first pulsar was discovered in 1967 by astronomer Jocelyn Bell Burnell, when she was a PhD student. Dr. Bell Burnell is highly esteemed for her trailblazing career as well as her commitment to diversifying representation in science; many people believe that she should have shared the Nobel Prize for Physics with her research advisor when he was given the award in 1974 for the discovery of pulsars (especially because this class of objects has greatly impacted the subsequent development of astronomy).

As pulsars spin, they give off regular pulses of light, which occur with extreme precision. But strangely, the timing of the pulses observed by Wolszczan and Frail was off. Pulsar PSR B1257+12, located 2,300 light-years away, should have been pulsing every 6.219 milliseconds, but every so often it missed the beat. After careful investigation, these astronomers came up with an explanation, the only one that made sense: the otherwise regular pulsing of PSR B1257+12 was being disturbed by *planets* in its orbit. There had to be at least two, each having masses at least three times that of Earth. They were, however, located much closer to their host star than we are to the Sun. The planets circled their star once every 67 and 98 days, similar to Mercury, which circles the Sun every 88 days. Wolszczan and Frail correctly predicted that there had to be a third planet, which was later found to have a mass no less than half of Mercury's and to complete an orbit every 25 days.

Three years after Wolszczan and Frail's discovery, the first detection of an exoplanet around a *Sun-like* star raised even more questions about the nature of extrasolar systems. This time, the system was a mere 50 light-years away, located in the Pegasus constellation. The Swiss astrophysicists Didier Queloz and Michel Mayor found a very massive exoplanet in close proximity to its parent star, 51 Pegasi. Dubbed 51 Pegasi b, it zips around its parent star every four days, which means that it is much closer to its parent star compared to Mercury and about five times hotter. Oddly, though, it was inferred to be much bigger and more massive than Mercury or any of the rocky planets in our Solar System. It has roughly half the mass of Jupiter and is about the same size. Because of its gigantic proportions and searing temperature, it was nicknamed "hot Jupiter." It was only the first of its kind to be discovered in the following years.

Before the discovery of 51 Pegasi b, astronomers didn't think such a planet was possible, because we don't have any like that in our Solar System. Today astronomers think they are quite common; one in ten stars in the Milky Way likely hosts a hot Jupiter. In 2019, Queloz and Mayor won the Nobel Prize for their discovery, which launched an entirely new field of study.

As of this writing, more than 5,000 exoplanets have been confirmed, and many belong to multi-planet solar systems. And how do you go about finding an exoplanet? Rarely can we see them directly, using telescopes. Planets are millions of times fainter than stars. Most of the time, they are completely outshined by their parent stars, like fireflies near a bonfire. So it is very inefficient to look for exoplanets using the methods we typically use to observe stars.

Direct imaging—taking pictures—is the least common method for finding and characterizing other worlds. Since stars tend to be so much brighter than their planets, this method is best suited to big planets sufficiently far from their host star. As of this writing, only about five dozen exoplanets have been discovered through this technique, which is fairly new. Direct imaging involves a device called a coronagraph, which can block light. Attached to the inside of a telescope, it obstructs the star's glare before it reaches the detector.

Direct imaging allows astronomers to actually see exoplanets, which lowers the chance of false positives. But for now, most exoplanets are discovered indirectly, by observing their influence on their parent stars. The majority have been located by means of two techniques: the *transit method* and the *radial velocity method*. For both of these methods to work, the planet-star system must have an edge-on orientation, as viewed from Earth. This is because

▶ **EXOPLANET ZOO** | The stunning digital image on the next pages visualizes 1,100 of the more than 5,000 worlds discovered since 1992. Rendered to scale, the planets increase in temperature, from left to right, as indicated by the gradation in colors. *IMAGE BY MARTIN VARGIC.*

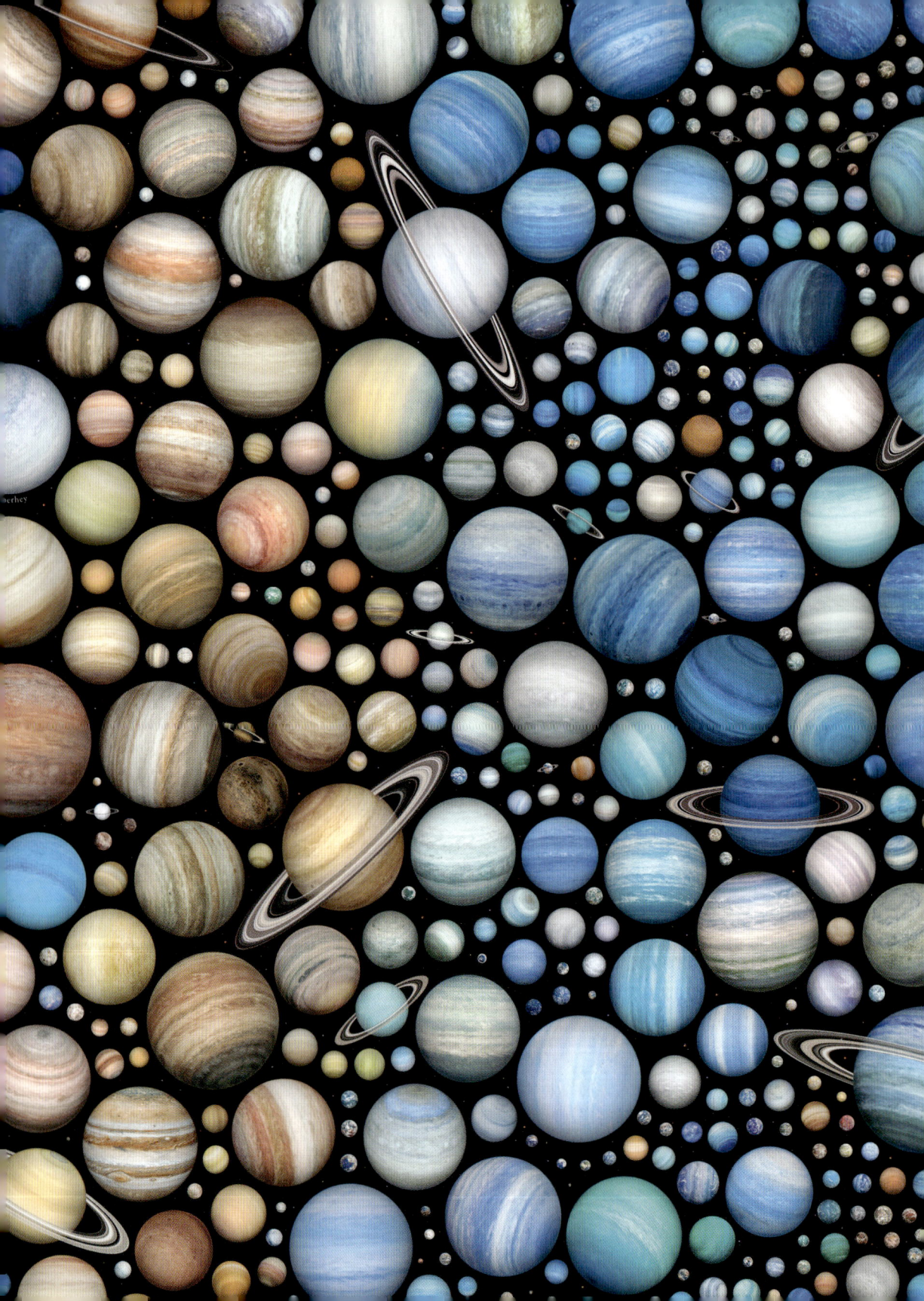

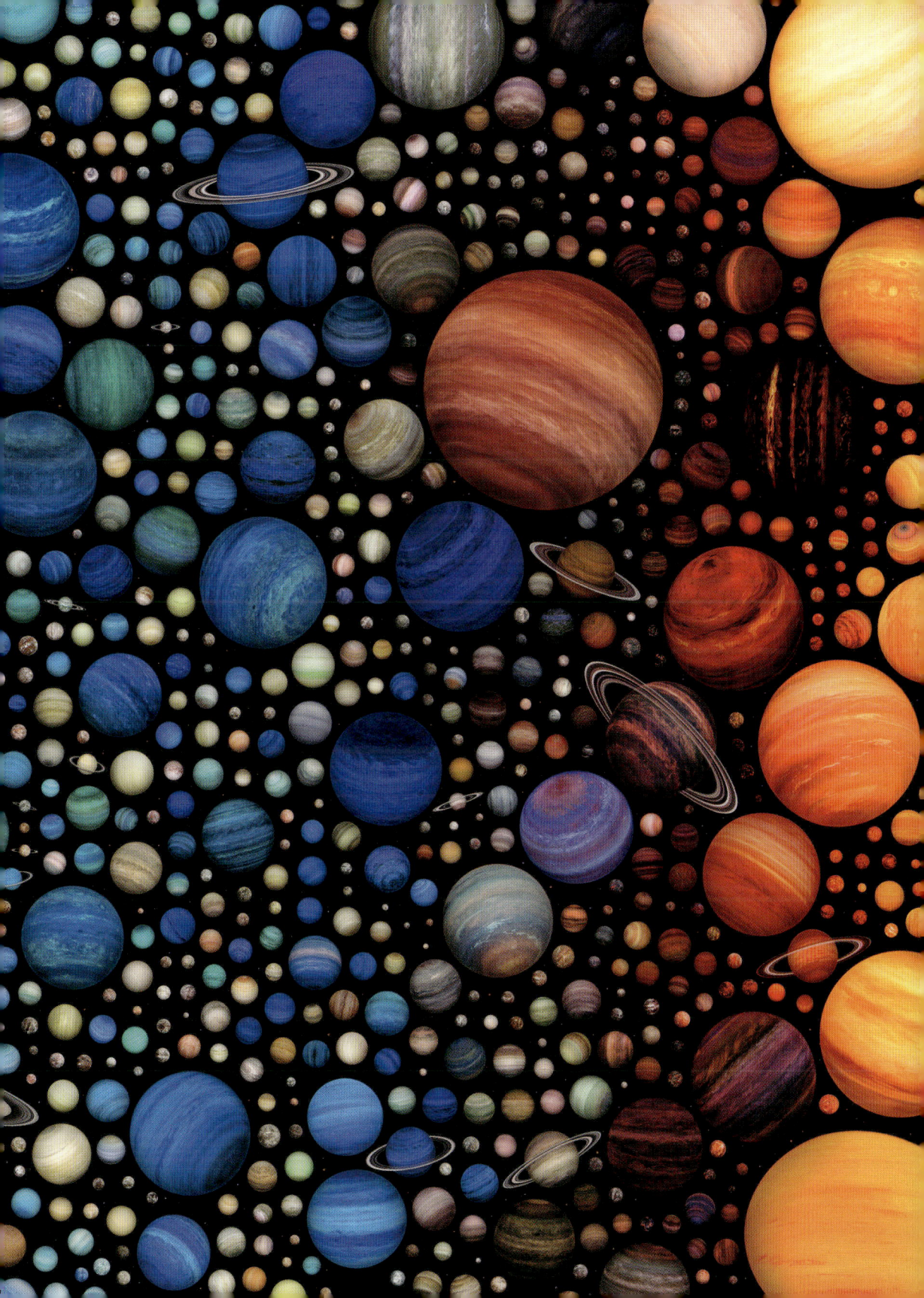

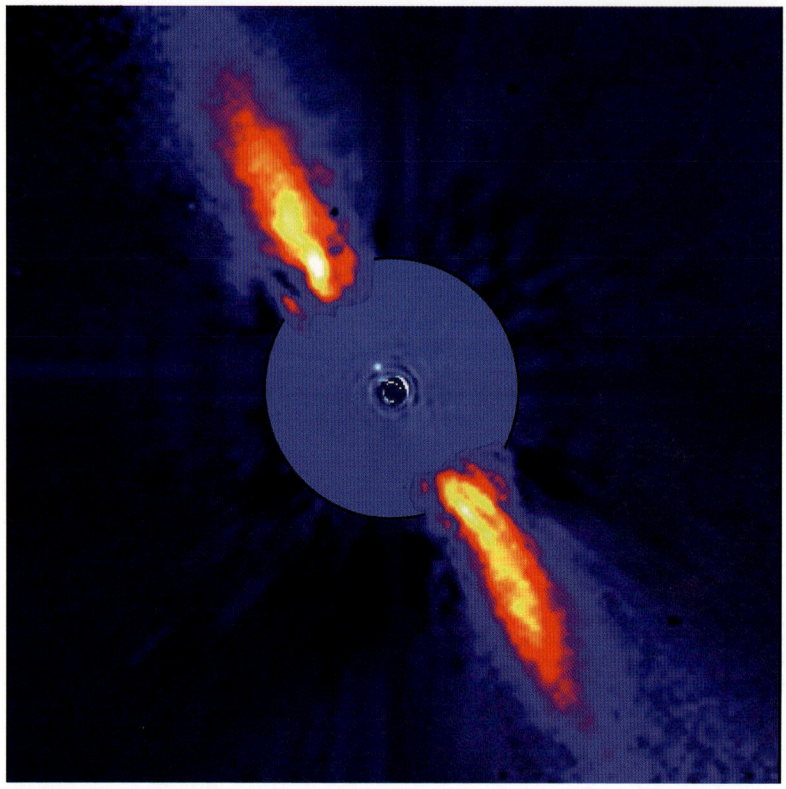

BETA PICTORIS B | Beta Pictoris b was discovered via direct imaging in 2008. The giant exoplanet is the little dot just upper-left of center. It is nearly 12 times more massive than Jupiter and circles its star once every 23.6 Earth years. What appears to be a pair of fiery wings is actually the star's debris disk. *Credit: ESO/A.-M. Lagrange et al.*

both techniques rely on astronomers being able to measure properties in the star's light that can be detected only when the planet-star system aligns with our perspective from Earth.

The *radial velocity method* depends on the fact that as a planet orbits a star, the star does not stay entirely still. In response to the gravitational tug of its planets, it wobbles around in a small ellipse. The more massive planets in an extrasolar system make the parent star wobble the most. In our Solar System, the point in the Sun around which the planets orbit—the barycenter—does not actually coincide with the exact center of the Sun. Jupiter and the other

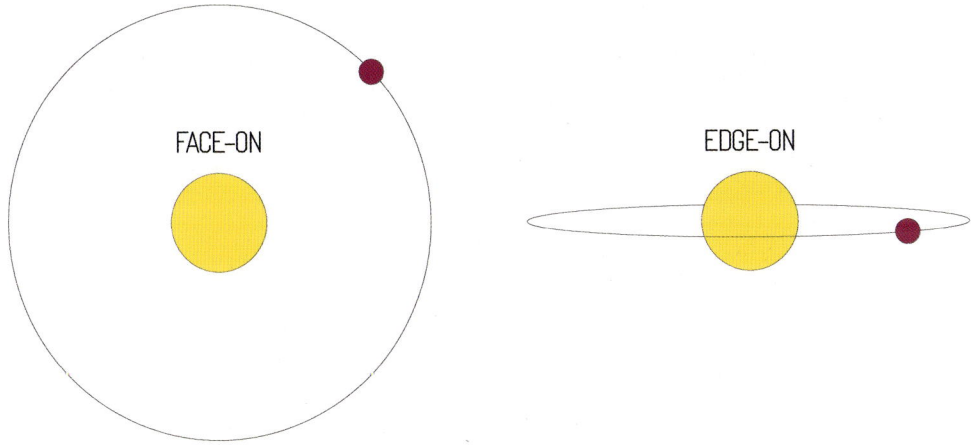

FACE-ON

EDGE-ON

gas giants cause the Sun to slowly wobble around a small ellipse once every 12 years. The less massive rocky planets—Mercury, Earth, Venus, and Mars—contribute a negligible amount to this effect.

When a star wobbles like this, if the system is aligned with Earth, the star repeatedly moves toward and away from us, and these slight motions—these changes in radial velocity—can be detected in the light it produces. As the star moves toward us, the wavelengths of light coming from it are compressed; when it moves away, they are stretched out. This shortening and lengthening of light waves due to motion is the same Doppler effect mentioned in Chapter 5. Current ground-based telescopes are sensitive enough to detect velocity shifts as small as about seven miles per hour (roughly three meters per second), the pace of a leisurely jog! More than 900 other planets, including 51 Pegasi b, were discovered with the radial velocity method. Still, it's challenging to apply in multi-planet systems. If several planets orbit a star, as is the case in our Solar System, it can be difficult or impossible to disentangle all of their contributions to the effect.

The most successful technique by far for detecting exoplanets is the *transit method*: When a planet passes in front of its star, it blocks some of the star's light. This results in a temporary dip in its brightness. Most of the time, the star's brightness remains roughly constant. But when a planet passes in front,

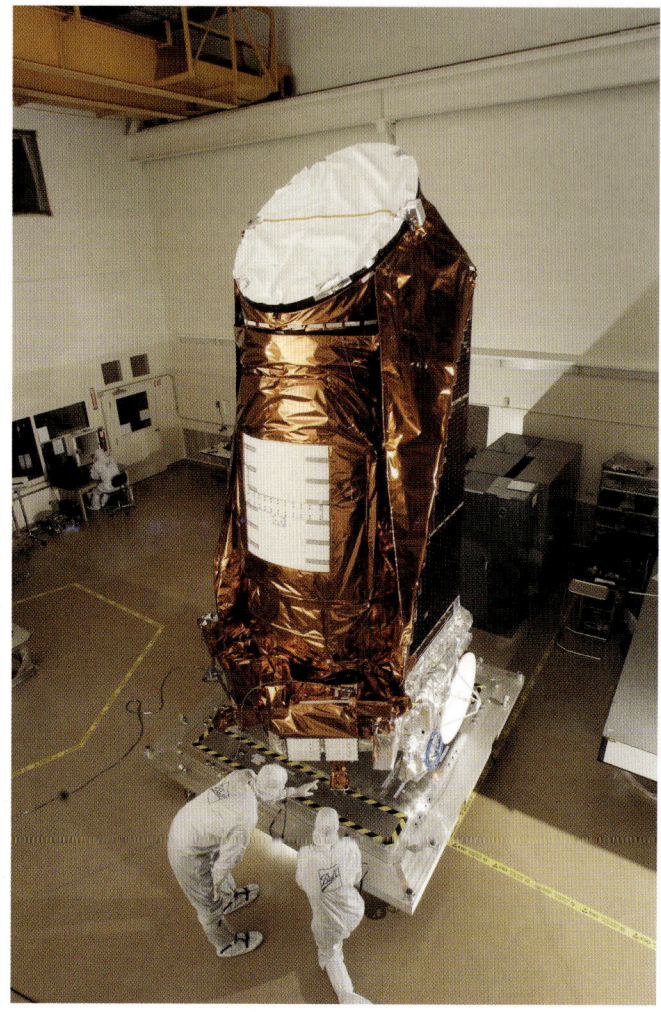

KEPLER SPACE TELESCOPE | The Kepler space telescope at Ball Aerospace & Technologies in Boulder, Colorado. About six months after this photo was taken, Kepler was launched on March 7, 2009.

its light is dimmed. Though scientists don't observe the planet directly, this method provides a lot of valuable information, such as the size of the planet and the distance between the planet and its star.

And that's not all. The transit method can also yield important information about the composition of an exoplanet's atmosphere. By observing how starlight filters through the atmospheres of planets, astronomers have been able to detect the presence of water vapor, carbon dioxide, methane, and other molecules in some of these atmospheres. The chemical composition of an exoplanet's atmosphere reveals details about its physical environment and whether it might be hospitable to life.

Astrobiology is an active area of exploration that includes the search for *biosignatures*, gases in exoplanet atmospheres that could arise as a byproduct of life. On our own planet, biosignature gases include oxygen, carbon dioxide, and methane. Launched in 2009, NASA's Kepler Mission was the most ambitious search for planets similar to Earth in size and mass. It used the transit method to find exoplanets and characterize their properties and those of their parent stars. For about nine years, the Kepler satellite surveyed a small region of the Milky Way, watching the shifts in brightness of more than half a million stars. Since most transits last no more than a few hours, the stars could be observed continuously, and repeated measurements were taken. By the end of the second phase of the mission, Kepler had confirmed the presence of 2,662 exoplanets, roughly two-thirds of all the known exoplanets at that time.

Like its eponym, the German astronomer Johannes Kepler, the Kepler Mission revolutionized our understanding of planets and of humanity's place in the universe, as it raised new questions about the possibility of life elsewhere in the cosmos. To begin with, the Kepler Mission showed that planets are common; in fact, they outnumber stars. Astronomers used the data to infer that every star in our Galaxy has at least one planet. Moreover, they could estimate that *one in four Sun-like stars is orbited by at least one Earthlike planet*! Given the number of Sun-like stars in our Galaxy, that puts the number at roughly 5 billion worlds in the Milky Way alone!

Another tantalizing result from Kepler is the sheer diversity of extrasolar systems. Of the thousands confirmed, most contain just a single planet. Fewer than 600 have two or more planets. Only one other system, called Kepler-90, has eight, like our Solar System. Another, the TRAPPIST-1 system, has seven. The host stars in many of these systems are cooler and smaller than the Sun, and their planets tend to be much closer to them. Another surprise from the Kepler Mission: Many exoplanets come in sizes unlike anything

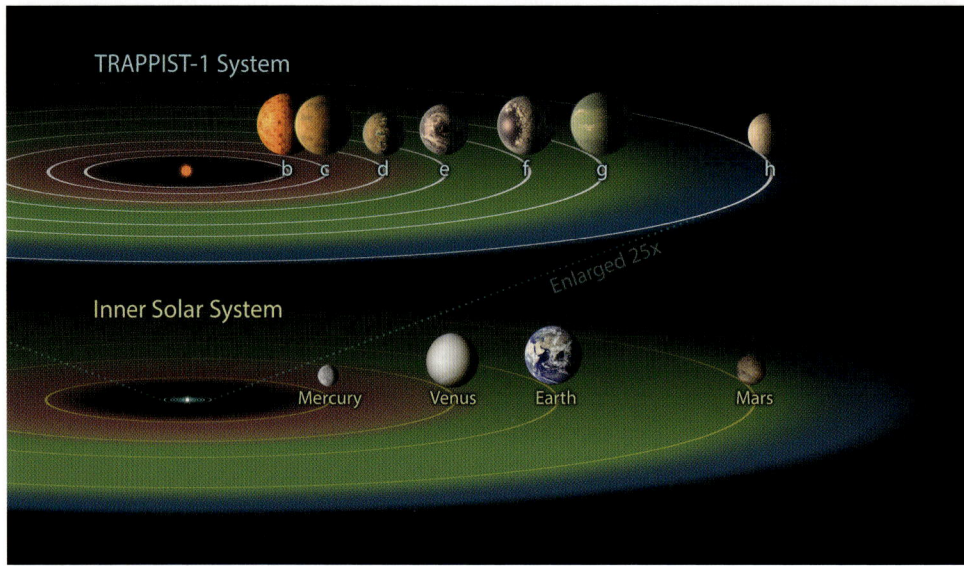

TRAPPIST-1 | The TRAPPIST-1 planetary system compared to the inner Solar System. The green shaded regions represent the habitable zones of each planetary system.

in our Solar System. They tend to fall in two distinct categories: "super-Earths," planets having radiuses as much as 1.5 times greater than Earth's, or "mini-Neptunes," planets from two to four times the size of Earth. Scientists are actively testing theories as to how these planets originated.

Kepler was designed to search for planets in the *habitable zone*, the region around a star where the temperature is warm enough for liquid water to exist on the surface of a planet. If a planet orbits too close or too far from its star, water may either evaporate or freeze. For this reason the habitable zone is often called the "Goldilocks zone." It cannot be too hot or too cold; it must be *just* right. Earth lies securely within the Sun's habitable zone; Venus and Mars are located near its inner and outer margins.

The TRAPPIST-1 system consists of seven planets orbiting a cool red dwarf star, the least massive kind of star, having a surface temperature less than half that of the Sun. Because the star is cooler, its habitable zone lies much closer in. The entire TRAPPIST-1 system could fit inside the orbit of Mercury, and four out of the seven planets reside within its habitable zone.

Perpetual Night, Crimson Rainbows

Two decades after the first exoplanet was found, exciting new discoveries have continued to pour in. In 2016, a team of astronomers conducted observations of Proxima Centauri, the star closest to the Sun, and confirmed that the anomalies in its radial velocity were caused by a planet orbiting in its habitable zone. At only 4.25 light-years away, this makes Proxima b the potentially habitable exoplanet closest to us. It's only a stone's throw away, astronomically speaking.

Soon after the discovery of Proxima b, two other planets were identified in this extrasolar system. Since they are located at distances much closer to and farther away from their host star, it is unclear whether they are habitable.

If life exists on Proxima b, what might it be like? How Earthlike is the planet? Because Proxima b is slightly more massive (1.3 times) than Earth, scientists believe that it may also have a rocky surface. Also like Earth, Proxima b sits squarely in its star's habitable zone. This means that if the planet has an atmosphere, its average temperature could be in the neighborhood of 85 degrees Fahrenheit (29 degrees Celsius), warm enough for liquid water. If there is no insulating atmosphere, however, temperatures on Proxima b could be as low as minus 40 degrees Fahrenheit.

What would a year on Proxima b be like? For one thing, it would be *much* shorter than an Earth year. Our planet is about 93 million miles from the Sun, while Proxima b is only 4.6 million miles from its star. Proxima b therefore circles its star much more quickly—one year on the planet lasts just 11.2 Earth days. Imagine, a birthday once every week and a half! If the planet has seasons, they might change every few days.

What would a day on Proxima b be like? Astronomers think that its rotation about its axis is synchronized with its revolution about its parent star. In other words, it takes the planet 11.2 Earth days to spin once around its axis, the exact time it takes to make one complete orbit around its star. Therefore one side of Proxima b always faces its star, just like one side of our Moon always faces Earth. One hemisphere experiences perpetual daylight,

▶ **LIFE ON PROXIMA B** | An artist's imagining of the surface of Proxima b. *Credit: ESO/M. Kornmesser.*

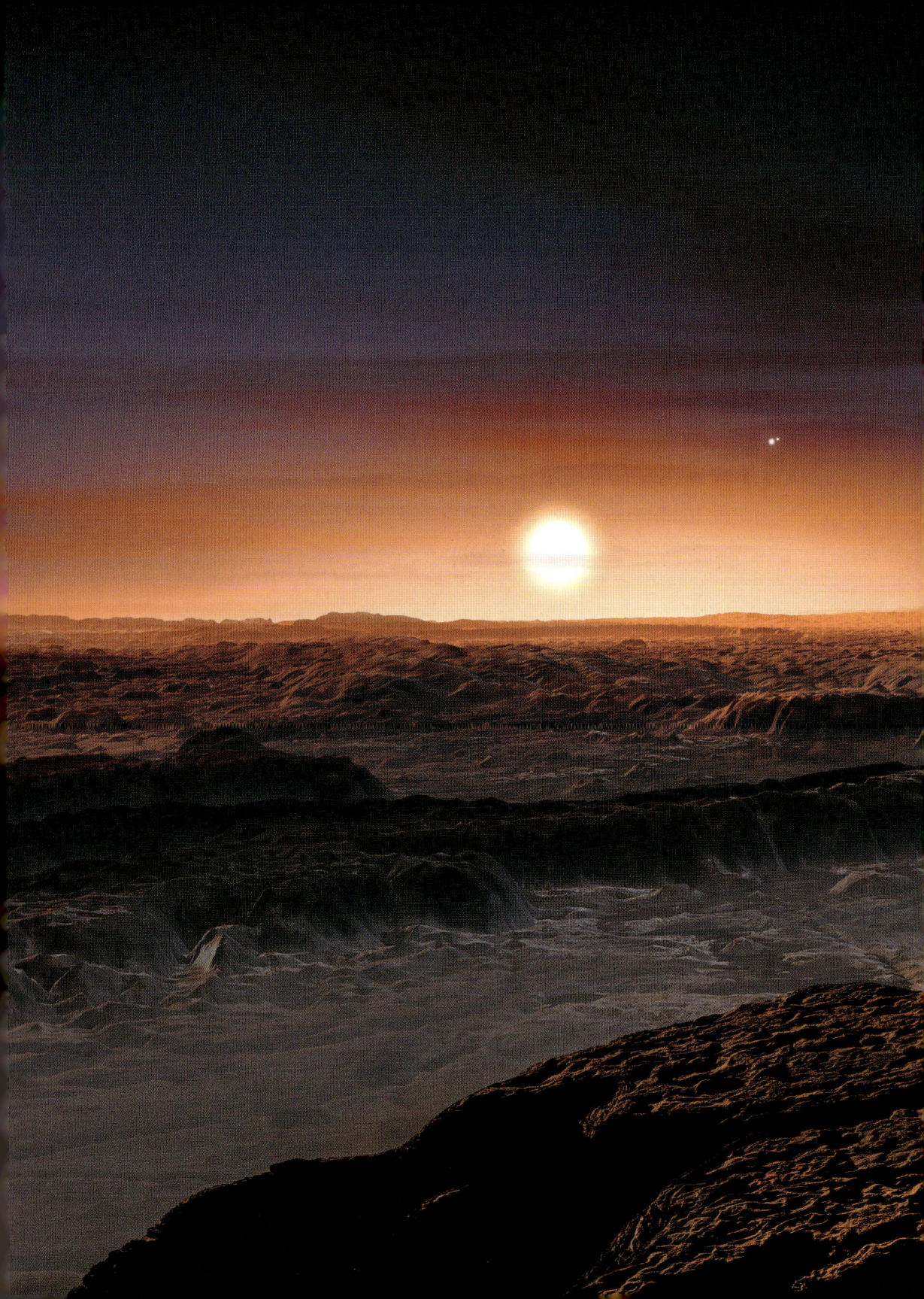

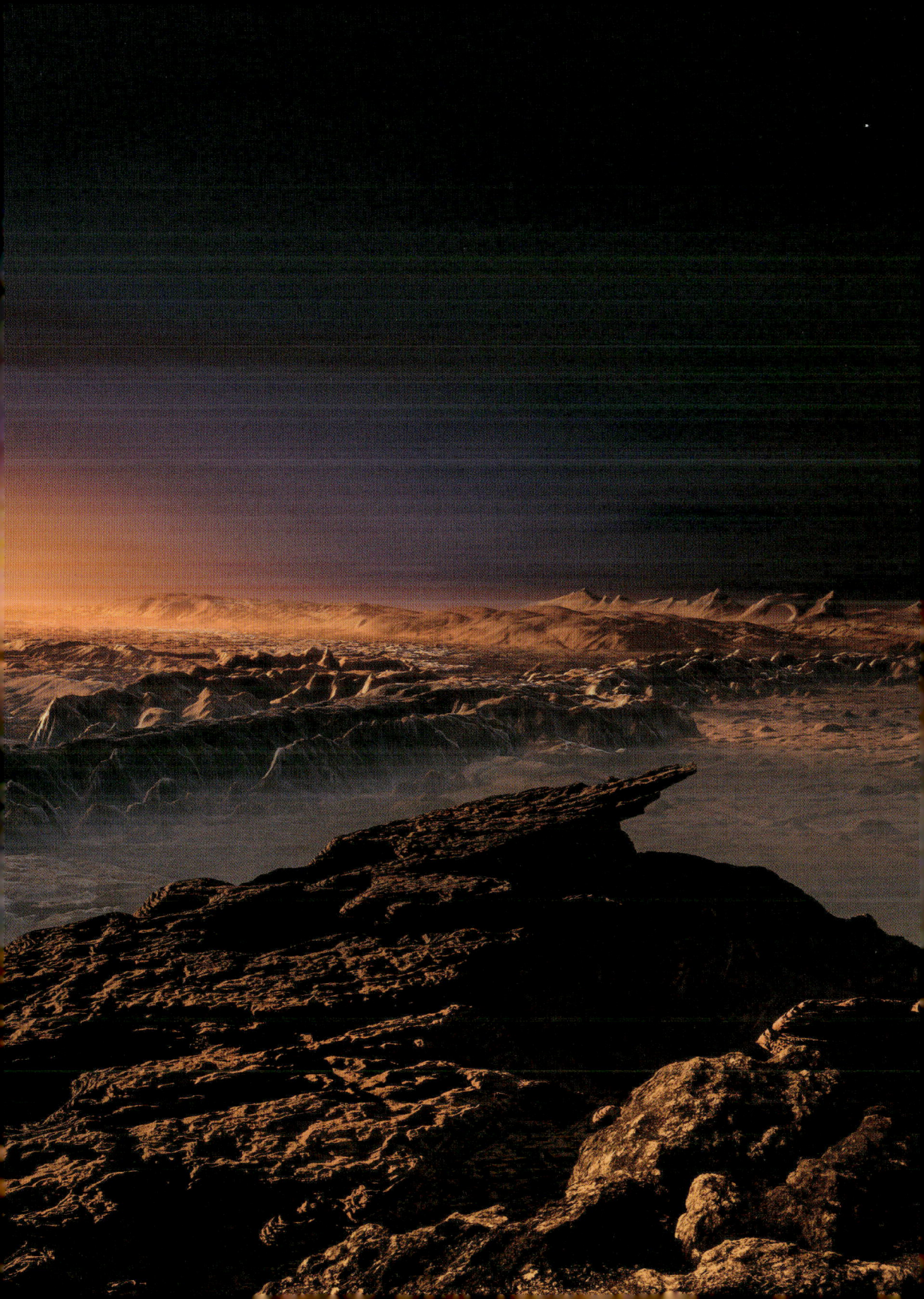

and the opposite hemisphere, perpetual night. This could have quite a strong effect on the global climate. It might be that the daytime side is much warmer on average than the nighttime side, which the planet's sun never lights up. But, supposing the planet has an atmosphere and weather, global air currents may transport energy from the dayside to the nightside, equalizing the overall temperature and making it possible for liquid water to exist on both hemispheres.

Perhaps there is a strip running around the planet, where the weather is stable enough for hypothetical Proxima beings to live. This could be sort of middle region between the dayside and nightside where our alien neighbors can experience the best of both hemispheres. Sadly, there are no sunrises or sunsets on Proxima b. But maybe this would be more than made up for by a magnificent view of the sky along this strip—the sun and stars always suspended in the vault together, on opposite horizons. As the planet orbits its sun, the constellations would constantly change. But from every location on the planet, the sun would maintain the same position on the horizon. For Proxima beings, perhaps the basic unit of time is not the "day," as we think of it here on Earth. Instead, maybe their astronomers developed a timekeeping system based on the cycles of the changing constellations.

What is Proxima b's sun like? Proxima Centauri is a red dwarf, the most common type of star in the Galaxy. This means that it is not only much cooler than our Sun, a yellow dwarf, but much smaller—about 1/10 the size of our star. Surprisingly, if we could stand on Proxima b, *its* sun would appear *larger* in the sky than ours does from Earth. That's because Proxima b is so much closer to its sun. For those living on the daytime side of the planet, there would always be a spectacular red orb hanging in the sky. If it rains there, Proxima beings might from time to time experience glorious crimson rainbows. That is, if their eyes have evolved to be like ours. If not, perhaps they see multicolored infrared rainbows, the likes of which we can only imagine. And while most Proxima beings may live on the daytime side, enjoying the balmy climate and perpetual sunshine, perhaps the astronomers live on the nighttime side where, without interruption, they can observe the stars to their hearts' content.

The "Ascent" of Man

Imagine turning on the news one morning to learn that humans had just made contact with ET. Would you be thrilled? Curious? Anxious? Relieved? According to a Gallup poll, more than one in three Americans believes life exists on Mars. A survey by the Pew Research Center found that two-thirds of Americans believe in the existence of *intelligent* alien life somewhere in the universe. Numerous polls have shown that belief in ET isn't exclusive to a particular culture or religion; large numbers of Christians, Muslims, Jews, and Hindus profess the belief that we have company in the cosmos. I sometimes wonder: If scientists discover solid evidence of ET, how much will it shake up the average person's worldview? How much would their life change?

As a professional stargazer, I love it when people ask me whether I think life exists elsewhere in the universe. It's a question close to my heart. My entry into real astronomy research was through SETI, the Search for Extraterrestrial Intelligence. One summer in the early 2000s, as an undergraduate physics major, I worked at the SETI Institute in Mountain View, California. My research project involved working with an array of radio telescopes called the Allen Telescope Array. This observatory, located a few hundred miles northeast of San Francisco, is dedicated to listening for radio signals from ET (and also conducts more traditional astronomical observations). One of my tasks was to characterize possible sources of human interference contaminating the signals we wanted to hear from ET. I worked with wonderful mentors, professional astronomers, at the SETI Institute. One of them was Dr. Jill Tarter, whose office I worked in that summer!

Dr. Tarter is one of the pioneers in the search for ET and a cofounder of the SETI Institute. She coined the term "brown dwarf," a celestial object often referred to as a "failed star," because at birth, it did not have enough mass to ignite nuclear fusion and become a star. A fellow alumna of UC Berkeley, Dr. Tarter is one of my heroines in astronomy.

When people ask me if I believe in ET, I tell them I'm agnostic. I don't know if life on Earth is the exception or the rule. I don't know how common life is in the universe. And however abundant life in the cosmos may be, the likelihood of *intelligent* life may be a different story. Simple organisms on Earth thrived for hundreds of millions of years before complex life emerged. Countless factors had to come together at the right times and under the right conditions in order for life—and eventually intelligent life—to get its foothold. Even if life in the universe is common, "intelligence" may not be.

Since my days as an undergrad, I've been fortunate to be involved in many other SETI research projects. I was part of the research team that in 2020 discovered the first exoplanet candidate in another galaxy. I truly consider the search for intelligent extraterrestrial life to be among the most worthwhile scientific pursuits. Yet as a scientist I think we should address this question in a methodical *scientific* way.

Science is a fundamentally experimental activity; it's concerned with things that can be measured, counted, empirically tested. The scientist tries to mold her curiosity into well-crafted questions that can be answered quantitatively. So instead of asking, "Are we alone?" an astronomer can ask, "What are the *odds* of finding ET?" In other words, if alien civilizations exist, how many are out there, and what are the chances that we will be able to detect them?

This question was first formulated by astronomer Frank Drake, a forerunner of modern SETI science and cofounder of the SETI Institute. Drake crafted a question that could address the main goal of SETI: to look for signals from technologically advanced civilizations that, like us, communicate with technology that relies on the transmission of signals, such as TV and radio.

Consider the reverse scenario—those aliens may be wondering if *we* exist. Humans have been broadcasting radio signals to communicate among ourselves for more than a century. These signals have been propagating through space ever since, at the speed of light. Clever alien astronomers who happen to be living on a planet within 100 light-years of Earth might be able to detect these signals, if their telescopes are sensitive enough. Traditionally, SETI science has focused its resources on "listening" for signals that could have unnatural origins in the *radio* portion of the electromagnetic spectrum.

At these long radio wavelengths, light travels unobstructed through vast reaches of space and is easily transmitted by Earth's atmosphere.

The first major telescope dedicated to listening for ET was Arecibo, the iconic radio dish in Puerto Rico that was once the largest of its kind in the world. The telescope was featured in the movie *Contact*, based on the novel by Carl Sagan, in which Jodie Foster's character—inspired by Jill Tarter—discovers evidence of ET. Today, the Allen Telescope Array, which consists of 42 radio dishes, operates as the first observatory specifically designed for SETI.

To get an idea of how many advanced alien civilizations we could communicate with in the Milky Way, Frank Drake thought about all of the factors that might be involved. To begin, we would have to know how many planets are in the Galaxy, how many of these are capable of supporting life as we know it, and which fraction might have evolved intelligent life. We would also need to know how long ET has been around on any such planets. Our intelligent ancestors thrived on Earth for some 6 million years before modern humans evolved in Africa. It would be another 200,000 years before we developed the technology to start communicating by means of radio signals. If ET is out there, it's unlikely that their stage of development would exactly match ours. They could be less technologically advanced or, perhaps, more sophisticated than we can imagine; to them, our cell phones and satellites would look primitive, the way the Stone Age tools of our ancestors look primitive to us.

Drake created an equation that takes these many considerations into account:

$$N = R_\star \times f_p \times n_s \times f_s \times f_i \times f_t \times L$$

Here, the number of technologically advanced civilizations in the Galaxy is called N. To calculate N, you multiply several other numbers. The tricky thing is, we *don't* know most of these numbers, though scientists do know some pretty well. For instance, astronomers are pretty sure that most stars have planets (largely because of the Kepler Mission), but we have no idea what fraction of alien civilizations have developed communications technology. The figure on the next page shows the definition of each quantity in the Drake equation and lists some commonly accepted values.

DRAKE EQUATION

$$N = R_* \times f_p \times n_s \times f_s \times f_i \times f_t \times L$$

N – Number of extraterrestrial civilizations that humans could communicate with

R_* – Rate at which stars are born in the Milky Way (1.5–3 stars per year)

f_p – Fraction of stars with planets (about 1)

n_s – Number of planets per solar system with conditions suitable for life (about 0.4)

f_s – Fraction of suitable planets actually harboring life (no idea!)

f_i – Fraction of life-bearing planets with intelligent life (no idea!)

f_t – Fraction of civilizations that have developed communications technologies (no idea!)

L – Length of time such civilizations have been broadcasting signals (no idea!)

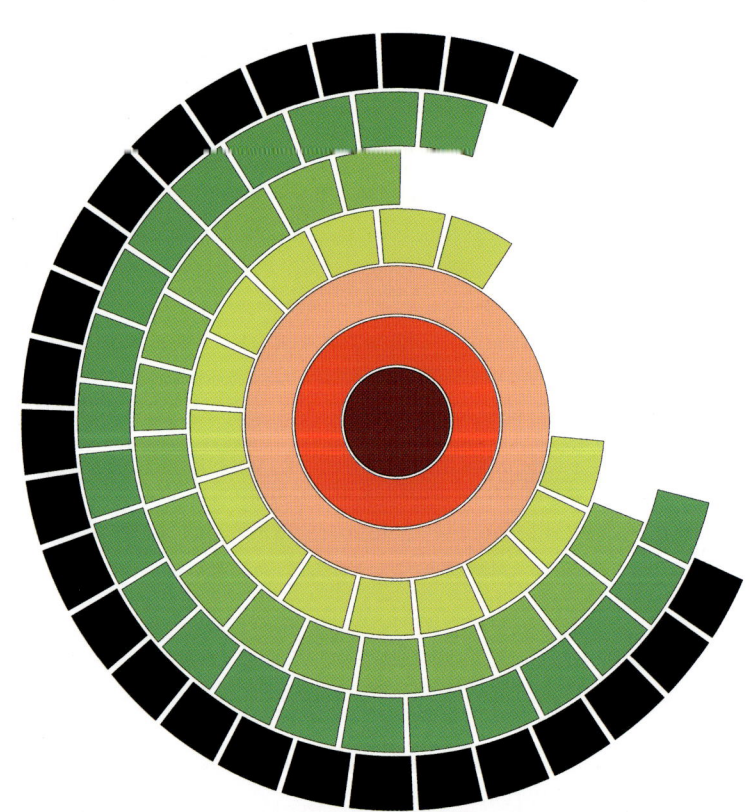

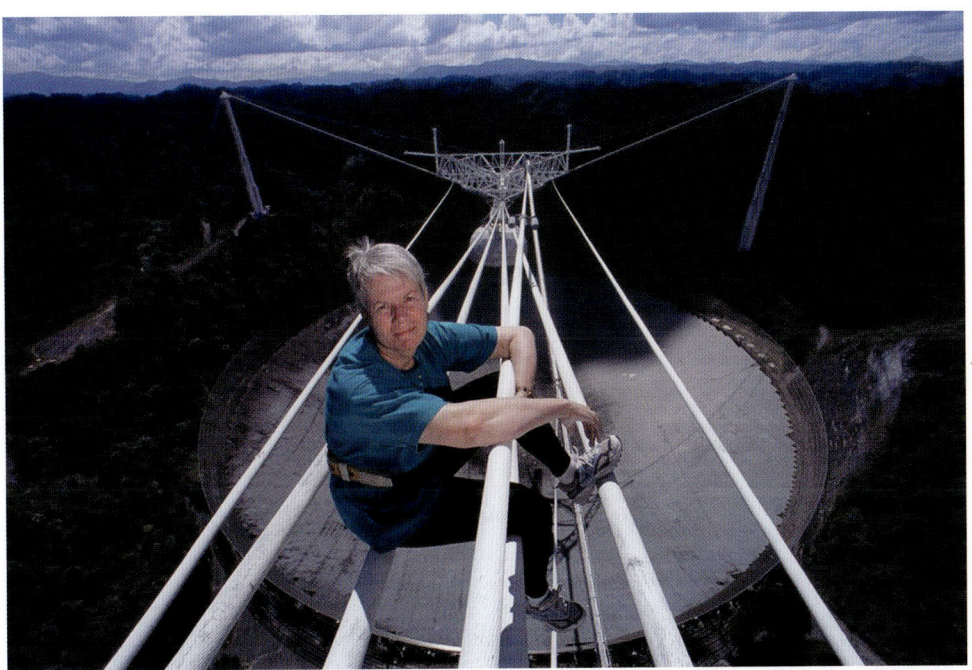

JILL TARTER | The cofounder of the SETI Institute, high above the Arecibo radio telescope in Puerto Rico.

Astronomers can measure the first three quantities with a high degree of confidence. The last four, however, are incredibly uncertain; we can only speculate. Depending on the values chosen for each variable, N can range from as low as 0.0000000001 (that is, we are most likely alone) to 16 million (guess what—the Milky Way is home to millions of civilizations). What a range!

It has been said that the Drake equation quantifies our ignorance, that it raises questions perhaps better left to philosophers and artists. Furthermore, what if the truly advanced civilizations out there wisely don't *want* to be found? Maybe they've devised ways to mask their signals. What if ET is listening in on *us*? Do *we* want to be found?

The universe is almost 14 billion years old. By comparison, humanity has been around for the blink of an eye. Perhaps numerous civilizations existed in the past but survived for only a short time before going extinct. The causes could have been natural, like the asteroid that slammed into Earth 66 million years ago and caused the dinosaurs to die out. Or maybe some civilizations self-destructed, wiped themselves out through war, pernicious widespread inequality, and violence to the environment. If *we* destroy ourselves before truly maturing as a species, we won't ever see the ancient extraterrestrial signals that may now be traveling from across the Galaxy toward our planet.

This raises yet another question: What do we mean by "advanced" and "civilization"? If we're honest, we must admit that greed, dishonesty, and violence characterize many aspects of human relationships between individuals, organizations, and nations. How "civilized" are we? "We misuse language and talk about the 'ascent' of man," Dr. Tarter once said. "We understand the basis for the interrelatedness of life, but our ego hasn't caught up yet."

If one day scientists do make contact with ET, what should happen next? Should we respond? Who would decide? Religious leaders, politicians, the privileged? Imagine we found that ET was living on an exoplanet a mere 100 light-years away. If we sent an encoded radio signal to make contact, it would take at least 200 years to receive a response. Suppose we could communicate with ET—what should we say? What would we ask them, for the sake of future generations?

Abena's Question

The human desire to discover life in the universe springs from a spiritual need to experience meaning, beauty, and mystery. Art emerges from the same need. Artists dare to give tangible form to the unseen realms and assert that there's more to life than meets the senses. The most powerful and enduring art is that which penetrates the surface things, dares to touch the heart and stir the soul.

In his essay "Concerning the Spiritual in Art," the modernist Russian painter Wassily Kandinsky speaks of "awakening subtler emotions" and the "emotional power of the artist." He believed that color can "directly influence the soul," that "the work of art is born of the artist in a mysterious and secret way. From him it gains life and being. Nor is its existence casual and

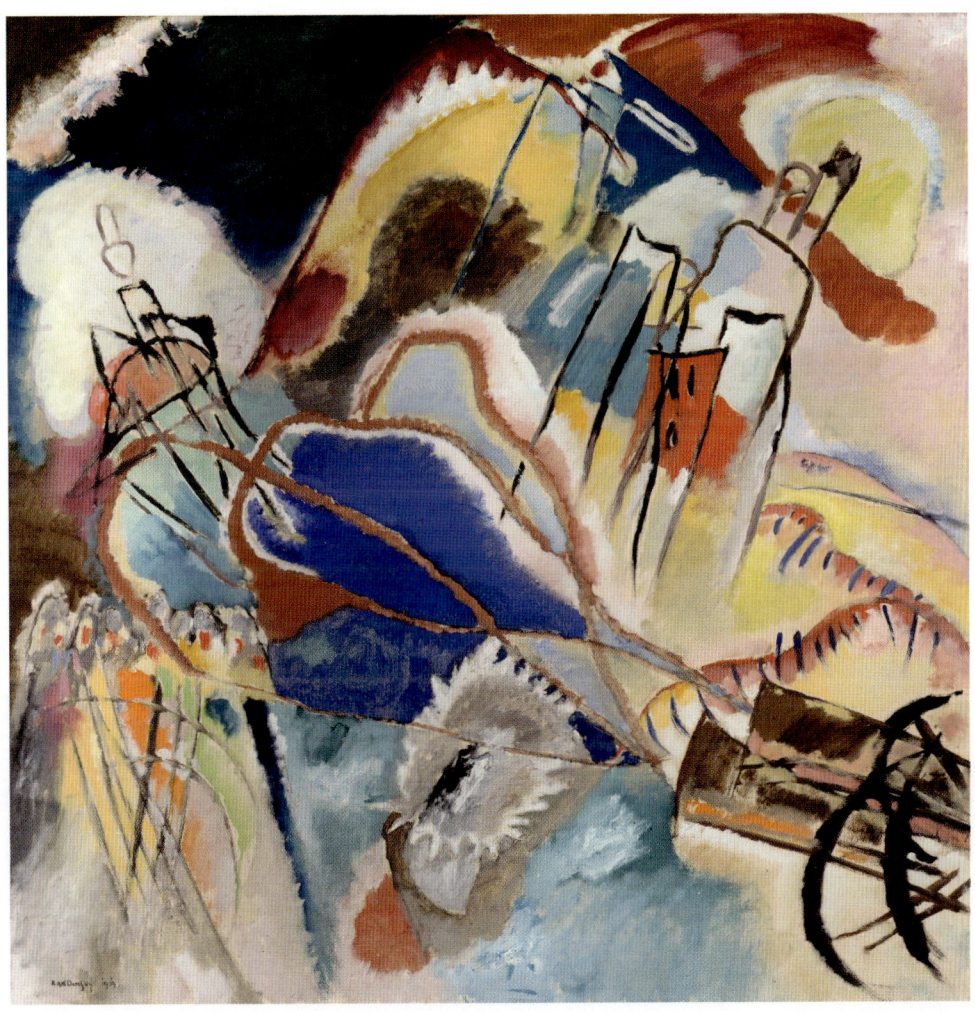

IMPROVISATION NO. 30 (CANNONS) | Wassily Kandinsky, 1913, oil on canvas, 43¹¹⁄₁₆ × 43¹³⁄₁₆ inches (111 × 111.3 cm). The Art Institute of Chicago.

inconsequent, but it has a definite and purposeful strength, alike in its material and spiritual life. It exists and has power to create spiritual atmosphere . . ."

Kandinsky's concern that materialism threatens to kill the most essential part of human life—spirituality—was reflected in paintings that became increasingly nonrepresentational and abstract over the course of his life. "The nightmare of materialism," he writes, "which has turned the life of the universe into an evil, useless game, is not yet past; it holds the awakening soul still in its grip." He believed in creating art that could encourage people to look past the outward appearance of things into their inner meaning.

What if we engaged in the work of science with the same discerning, spiritually sensitive attitude that Kandinsky urges us to adopt as we engage with works of art? Art is spiritual food. Science also has the potential to nourish needs other than the purely physical and intellectual.

I've been asked dozens of times whether I think aliens exist. One of these times followed a presentation I gave about the cosmos to a group of schoolchildren. I met Abena in Chorkor, a seaside community of Accra, Ghana, where many of the people fish to make a living. Abena wore a school uniform and a neat, closely cropped haircut like the other children. After my talk, she stood up and asked a question in a barely audible voice. Mic in hand, she whispered, "Are there people living on other planets?"

I smiled, admiring her because she was more brave than shy. Clearly, this was a pressing question for her. Instead of answering her directly, I asked what *she* thought. Shyness seemed to repossess her. Half shrugging, she murmured something I couldn't hear and sat back down.

Since that day in Chorkor, my imagination comes alive when I think about Abena's question, about why it might have been important to her. When Abena daydreams about people on other planets, I wonder, does she imagine them to be like her and the people in her community, where they make their living catching and smoking fish? Are they poor like her people? After school, do those extraterrestrial boys and girls go home to shacks without electricity or running water? Are they black like her, or white like the Jesus in the cheap picture frames peddled in downtown Accra, or some other color? Is survival a constant struggle for them, too?

And I'm fascinated by a word in Abena's question: "Are there *people* living on other planets?" Not aliens. People. In her mind, as she pictures these people—inhabiting different kinds of bodies, eating different kinds of food—does she imagine them having different values, too? What would she have thought if I had told her *Yes, there are people living on other planets?* How would that revelation have impacted her life?

I believe children start life inherently filled with a sense of mystery and naturally lean toward joy and hope. If Abena received an affirmative answer to her question, I imagine the news might have delivered to her spirit a fresh surge of all three. Indeed, it is a hopeful, joyful, mysterious thing to imagine people living in unknown parts of the universe. Living lives that are perhaps more peaceful, civilized, loving, and grounded in the confidence of the overwhelming abundance of life.

This gets to the heart of why the search for life is so significant. Making contact with ET is a grand ambition, and there are no guarantees that we will accomplish it. But the ultimate value of pursuing a huge goal outside oneself is not the attainment of the goal itself; the value is that one must rise higher in the process of aiming for it. No matter the outcome, the quest to discover life in the universe is a worthy pursuit if we engage in it in a way that fosters hope, creates joy, and cultivates mystery.

Life beyond Earth may look nothing like what we're familiar with. Perhaps our ability to recognize it in an alien form will depend on our ability to engage with science with a more spiritually sensitive attitude. What would that even look like?

We may not know what life is exactly, but we each have a personal sense of what makes us feel alive. We all know the feeling of being filled with life and of being drained of it. A spiritually sensitive posture toward science rests on the same foundation as a spiritually sensitive life—love. We feel most alive, like our lives have meaning, when we love and serve others.

Western society is very individualistic, and it seems we live in a time when people feel more disconnected and alone than ever. The materialism condemned by Kandinsky and the associated egotism called out by Tarter are major causes of this epidemic of alienation. Einstein made a related observation. In an article written almost three decades after winning the Nobel Prize, he observed: "All human beings, whatever their position in society, are suffering from this process of deterioration. Unknowingly prisoners of their own egotism, they feel insecure, lonely, and deprived of the naïve, simple, and unsophisticated enjoyment of life. Man can find meaning in life, short and perilous as it is, only through devoting himself to society."

The brilliant scientist and artist George Washington Carver identified the same maladies (materialism, egotism) and the remedy (service): "It is not the style of clothes one wears, neither the kind of automobile one drives nor the amount of money one has in the bank, that counts. These mean nothing. It is simply service that measures success."

A devotion to service to others, to reaching out to others, makes us feel alive and creates a sense of meaning that can't be had in any other way. Service enlarges us. We feel less "alienated" from our fellow humans and more deeply connected. We feel hope.

At this point, I recall some words of the young Fyodor Dostoevsky, penned in a letter to his brother years before he would conceive his great works, *Crime and Punishment* and *The Brothers Karamazov*. Dostoevsky wrote the letter on a fiercely cold day in December 1849, a few hours after he had suffered through the trauma of a mock execution. Just 28 years old, he was being transported to Siberia, where he would be imprisoned and exiled for the next eight years of his life. Out of this moment of trauma, this is what he wrote: "Brother, I'm not depressed and haven't lost spirit. *Life everywhere is life, life in ourselves and not in the external. There will be people near me, and to be a human among human beings, and remain one forever,** no matter what misfortunes befall, not to become depressed, and not to falter—this is what life is, herein lies its task."

Dostoevsky's amazing words foreshadow those of his fellow Russian, Kandinsky. "*To be a human among human beings* . . . this is what life is."

The sky proliferates with beautiful mystery. When I look up, I have hope in humankind's potential for transformation, because whether alien life exists

---

* My emphasis.

on other planets or only in our dreams, life is everywhere around us. Life responds to life, and only a spark of it is needed to kindle an explosive chain reaction of hope.

I don't know whether aliens exist. Whatever we end up finding, there's one thing I'm confident of. This knowledge, and its implications for us, will not outweigh the immutable responsibility we already have toward one another. If it happens that life exists nowhere else in the universe but on Earth (though a negative cannot be proven), we may conclude that life is supremely precious, and we ought to learn to love one another. If, on the other hand, scientists were to accumulate irrefutable evidence that we are *not* alone, we would have further scientific proof that we're fundamentally connected to one another and to the cosmos in a profound way and that we should, therefore, love one another.

CHAPTER SIX

SPACE—TOUCHING THE INVISIBLE

*The aim of art is to create space—space that is not compromised
by decoration or illustration, space within which the subjects
of painting can live.*
— *Frank Stella* —

Space is the breath of art.
— *Frank Lloyd Wright* —

*Some people say they feel very small when they think about space. I felt
more expansive, very connected to the universe.*
— *Mae Jemison* —

Everything in the universe is in motion, including space itself. The movement of space is invisible, but if we *listen* closely, we can hear its songs resonating from across the cosmos.

Two black holes dance together in some remote region of the universe. They circle each other, at first slowly. But the tempo gradually increases as their gravitational attraction pulls them closer. As they spin, they make their own music, flinging out vibrations through space that can't be seen. They draw closer and closer together, irresistibly, until suddenly, they merge. They release one final burst of song, their coda, the ultimate embrace. Billions of years later, faint echoes of their dark song arrive here on Earth.

For most of history, astronomy was the science of vision and light. Astronomers over the centuries constructed a story of the cosmos by decoding the messages that light carries on its long journey through space. This all changed in 2015, when scientists first detected ripples in the ocean of spacetime.

Ripples in Spacetime

We're used to thinking in terms of three-dimensional *space*. But scientists think in terms of *spacetime*, a four-dimensional model of the universe that links space *and* time. Albert Einstein demonstrated that space and time are, in fact, one and the same. Saying "spacetime" is simply a physicist's way of saying "the universe." The concept of spacetime is useful because it helps to explain various phenomena we observe in the cosmos, including gravity. Einstein showed that spacetime is malleable; it can curve and bend. He showed that gravity is not a "pulling" force but rather the actual curvature of spacetime, the curvature of the universe itself. This is one of the key ideas of his theory of general relativity.

What causes the curvature of the universe? The presence of massive bodies. Anything that has mass curves the space around it. Imagine putting a baseball on a stretched sheet. The ball causes the fabric to bend around it, and any other object placed nearby—a marble, for instance—will fall toward it. This is how gravity works. Massive bodies cause spacetime to curve so that other objects fall toward them. In this analogy, the fabric represents spacetime (the universe), the baseball could represent a star, and the marble could represent a planet. But anything that has mass, including you and me, causes spacetime to curve around it. More massive things cause greater curvature, like a bowling ball placed on a sheet causes it to bend more than a marble.

What happens when massive bodies move in a certain way? Einstein predicted that they send out ripples in the fabric of spacetime, which he called *gravitational waves*. These waves radiate from their source at the speed of light. A century after he prophesied their existence in 1915, scientists at last reported the first direct detection of gravitational waves, in effect unlocking a new sense with which we could perceive the universe.

The cause of the ripples detected in 2015 was the collision of two black holes located 1.3 billion light-years from Earth. This event was the first observation of gravitational waves *and* the first direct observation of black holes. Black holes are "black" because they do not emit any light at all, making it impossible to observe them directly using telescopes. Before this discovery, astronomers could "see" black holes only indirectly, by observing their effects on the stars and glowing gas in their immediate surroundings.

Gravitational waves are generated by entirely different physical mechanisms than light, and special instruments are required to sense them. The story of the Laser Interferometer Gravitational-Wave Observatory (LIGO) began in 1967 when Rainer Weiss, then teaching at MIT, conceived a version of the idea for a classroom exercise.* Built in the 1990s, LIGO consists of two laser interferometers, one located at Livingston Observatory in Louisiana and the other at Hanford Observatory in Washington State. These strange-looking observatories have long arms, each 2.5 miles (4 kilometers) in length, positioned at right angles. Laser beams are fired back and forth along inside vacuum chambers buried underground beneath the arms, at the intersection of which are mirrors. Gravitational waves passing through, as they squeeze and stretch space, cause the distances between the mirrors to change. The two LIGO detectors, operating in unison, are designed to detect unbelievably minute distortions in spacetime less than 10,000 times the diameter of a proton!

---

* An interferometer is a group of telescopes acting as one.

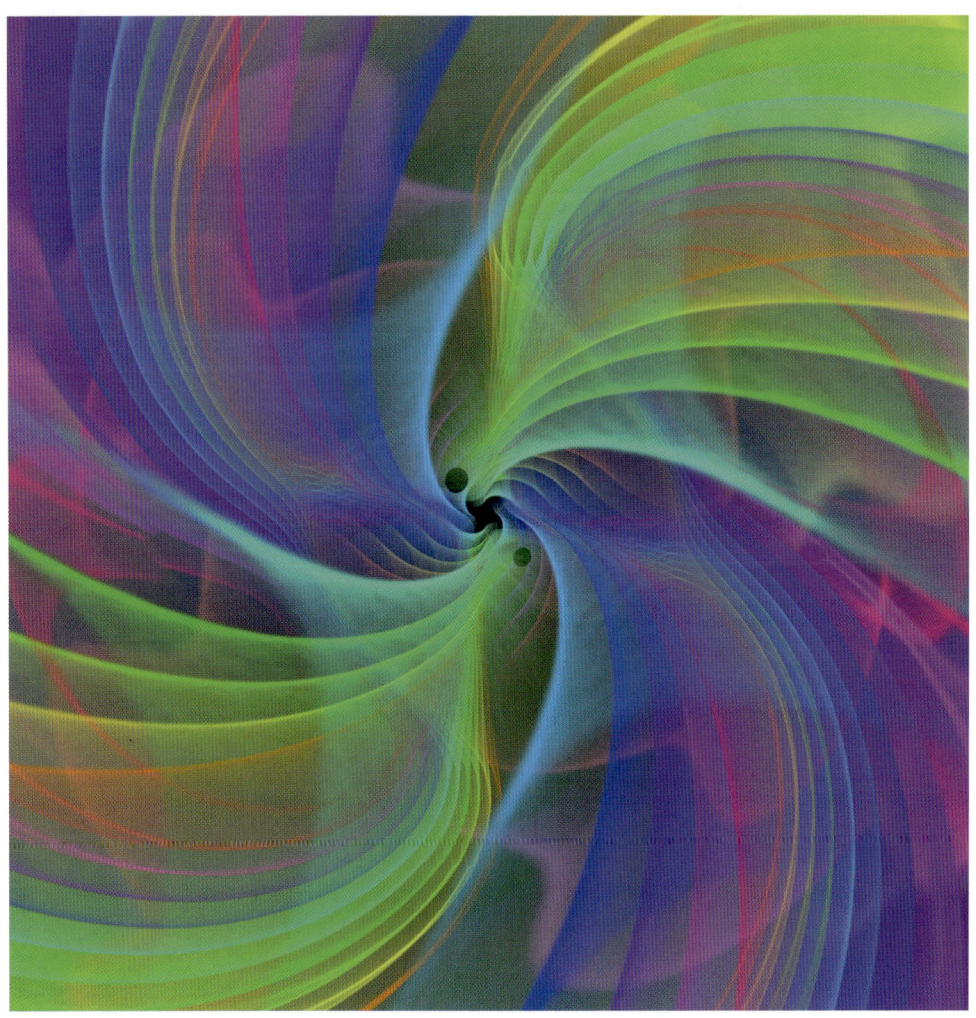

GRAVITATIONAL WAVES | A scientific visualization of gravitational waves rippling through spacetime.

When the binary black holes discovered by LIGO merged, they momentarily released more energy in the form of gravitational waves than is contained in all the stars in the universe! But this event lasted only an instant. The waves dissipated as they rippled through space, and by the time they reached us 1.3 billion years later, in 2015, the amount of spacetime "wiggling" they produced was *1/1000* the size of an atomic nucleus.

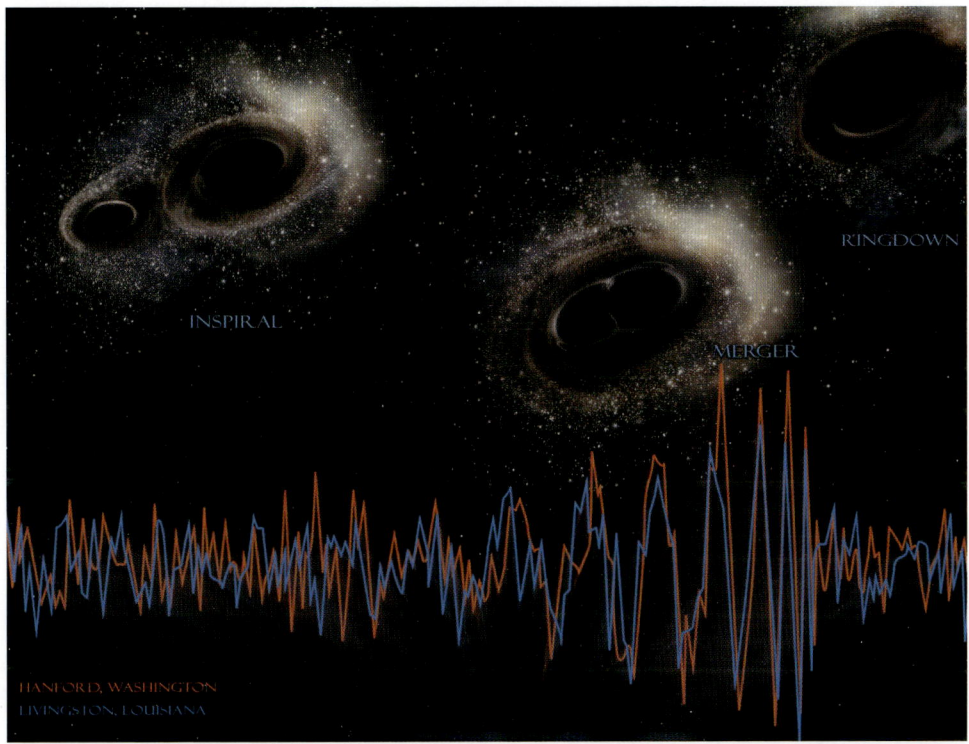

FIRST BINARY BLACK HOLE MERGER DETECTED BY LIGO | The figure shows the signals detected at the two LIGO observatories during the inspiral, merger, and ringdown stages of the event.

As the black holes spiraled around each other, the gravitational waves they released carried away some of their energy. As a result, they moved closer together and spiraled faster, which caused the gravitational waves to become more intense and carry off even more energy. This inward spiraling (inspiral) produced a signal that was detectable for a fraction of a second. As the pair drew even closer, the signal evolved into a sudden "chirp," followed by a "ringdown" of the final, single black hole.*

---

* The terms "chirp" and "ringdown" refer to the sounds we would hear from the black holes if we transposed the original frequency of their vibrations up to the range of human hearing.

Scientists could measure the properties of the black holes by comparing the signal observed by LIGO with the theories they had developed about the way they merge. They discovered that the masses of the black holes were 29 and 36 times that of the Sun (that is, 29 and 36 solar masses). The final black hole measured about 62 solar masses. That means three solar masses had gone missing. Where did they go? They were radiated away as energy in the form of gravitational waves!

Since the 2015 report of the binary black hole collision, LIGO has detected several more such events, as well as merging pairs of neutron stars. After a very massive star explodes as a supernova, it leaves behind a neutron star, the densest, tiniest star known. A neutron star may have a solar mass worth of material crammed into a ball just 12 miles across, so that a teaspoon of the stuff could weigh a billion tons!

Unlike black holes, neutron stars are luminous. This means that when they collide, they also release electromagnetic radiation. In 2017, the first neutron star merger was observed in both gravitational waves and light. The gravitational waves were detected by LIGO and Virgo, the European counterpart to LIGO located in Italy. Less than two seconds later, a burst of gamma rays was observed by Fermi and INTEGRAL, two space telescopes specially designed to observe this type of high-energy radiation. Analysis of the light-based observations revealed that collisions like this produce heavy elements, including lead and gold, which are flung into space by the force of the blast.

Observations of gravitational waves from colliding black holes and neutron stars provided further evidence for the veracity of Einstein's theory of gravity, described earlier. Yet the observations revealed a new mystery. How did some of these black holes form? Before LIGO began detecting black holes, astronomers knew about two distinct types: stellar-mass black holes and supermassive black holes. Stellar-mass black holes result when the core of a star much more massive than the Sun collapses under the weight of its own gravity, while the rest of it explodes as a supernova. The black hole left behind by a star of 130 solar masses could have a maximum mass of about 65 solar masses. Stars more massive than that, up to a certain limit, are expected to explode so powerfully that they leave behind nothing at all. Supermassive black holes are formed by entirely different mechanisms. They are found at the centers of many galaxies, including the Milky Way, and measure millions or billions of solar masses.

Most of the black holes observed by LIGO fall into the range of 20 to 50 solar masses—these are stellar-mass black holes. But in 2019, a new discovery was made—the most massive pair of black holes to date. They had masses 66 and 85 times that of the Sun, and when they merged, a 142-solar-mass black hole was formed. It falls squarely in the gap between stellar-mass and supermassive black holes. Thus scientists observed a new class of *intermediate-mass* black hole binaries, one that cannot be explained by our best models of stellar evolution. Astronomers are still puzzling over how they could have formed. Some researchers have suggested that each was built up from the merger of multiple black holes. Another possibility: The signal interpreted as the presence of intermediate-mass black hole binaries wasn't produced by black holes at all. Instead, it represented an unexplained phenomenon never before encountered.

This example illustrates one of the most wonderful things about the process of science: when done "right," it will always raise more questions than it answers.

Seeing the Invisible

A black hole is a region of spacetime where gravity is so strong, nothing can escape it, not even light! A black hole is born when a massive body, such as a star, is compressed so much that the speed needed to leave its gravitational field becomes equal to the speed of light.

Whether a star evolves into a stellar-mass black hole mostly depends on the mass it starts out with at birth. After a star with more mass than about 8 to 10 solar masses runs out of fuel in its core, it will explode as a supernova, leaving behind a small, compact remnant. In most cases, the compact remnant will be a neutron star with a mass of about 1.4 times that of the Sun. But if the initial star was at least about 20 solar masses, the compact remnant will consist of anywhere from about three to tens of solar masses. We expect that a star with more than 200 solar masses will not become a supernova but instead collapse straight into a black hole. (We'll discuss how supermassive black holes may form shortly.)

Whatever its origins, all of the mass in a black hole is compressed into an infinitely tiny point called a *singularity*. And its properties can be summarized with just three numbers: its mass, spin, and electrical charge. Given this simplicity, John Wheeler—the astronomer who coined the term "black hole"—famously said that "black holes have no hair." The theoretical physicist Subrahmanyan Chandrasekhar expressed their simple elegance in this way: "The black holes of nature are the most perfect macroscopic objects there are in the universe: the only elements in their construction are our concepts of space and time."

Simplicity notwithstanding, "our concepts of space and time" completely disintegrate at the singularity of a black hole. What happens when an enormous amount of mass is trapped in an infinitely small point? Gravity becomes infinite, and therefore spacetime curves infinitely. In other words, the known laws of nature shut down.

Nothing that gets swallowed up by a black hole has ever come back to tell us what happens inside. Fortunately, Earth itself will never become a black hole. Given its small mass, the entire planet would have to miraculously be stuffed inside a tiny region just 0.35 inches (9 millimeters) across. The mass of a black hole (or a potential black hole) is what determines its "size"—the distance between its center and *event horizon*, the boundary beyond which light cannot escape. As a black hole swallows up material (like stars or gas clouds) that happens to cross the event horizon, it will grow in mass and expand in size. But Earth is not gaining mass, and there's no physical mechanism that would cause it to implode. Neither will the Sun ever become a black hole. It would have to collapse into a region less than two miles (3.2 kilometers) across, more than 200,000 times smaller than its current size.

If black holes don't emit light, how did we know, before LIGO, that they really exist? Stellar-mass black holes are typically observed in binary systems that include a black hole and a normal star. As they orbit each other due to their mutual gravity, the black hole can strip gas from the surface of its

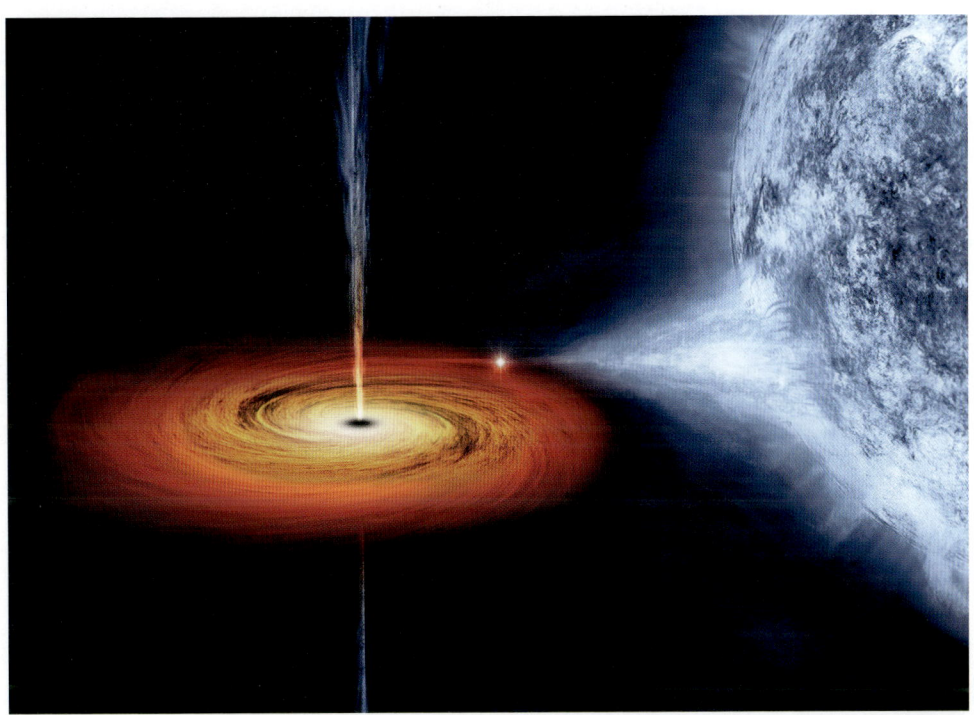

A STELLAR MASS BLACK HOLE | Artist's representation of a stellar-mass black hole in a binary system.

companion. As the outer layers of the star get pulled in toward the black hole, the gas accumulates into a rapidly spinning disk before eventually falling in or being deflected. The gas in the disk heats up so much that its x-ray glow may be observed across vast distances. This is how the IC 10 X-1 black hole was first observed. Located in a small galaxy 1.8 million light-years away, IC 10 X-1 weighs in at 23 to 33 solar masses, making it one of the most massive stellar-mass black holes known in the local universe.

The supermassive black holes lurking in the middle of most large galaxies are also observed via indirect methods that allow us to infer their mass. Sagittarius A*, the supermassive black hole at the heart of the Milky Way, is

▶ **CENTER OF THE MILKY WAY** | The center of our Galaxy at radio wavelengths, as observed by the MeerKAT radio telescope.

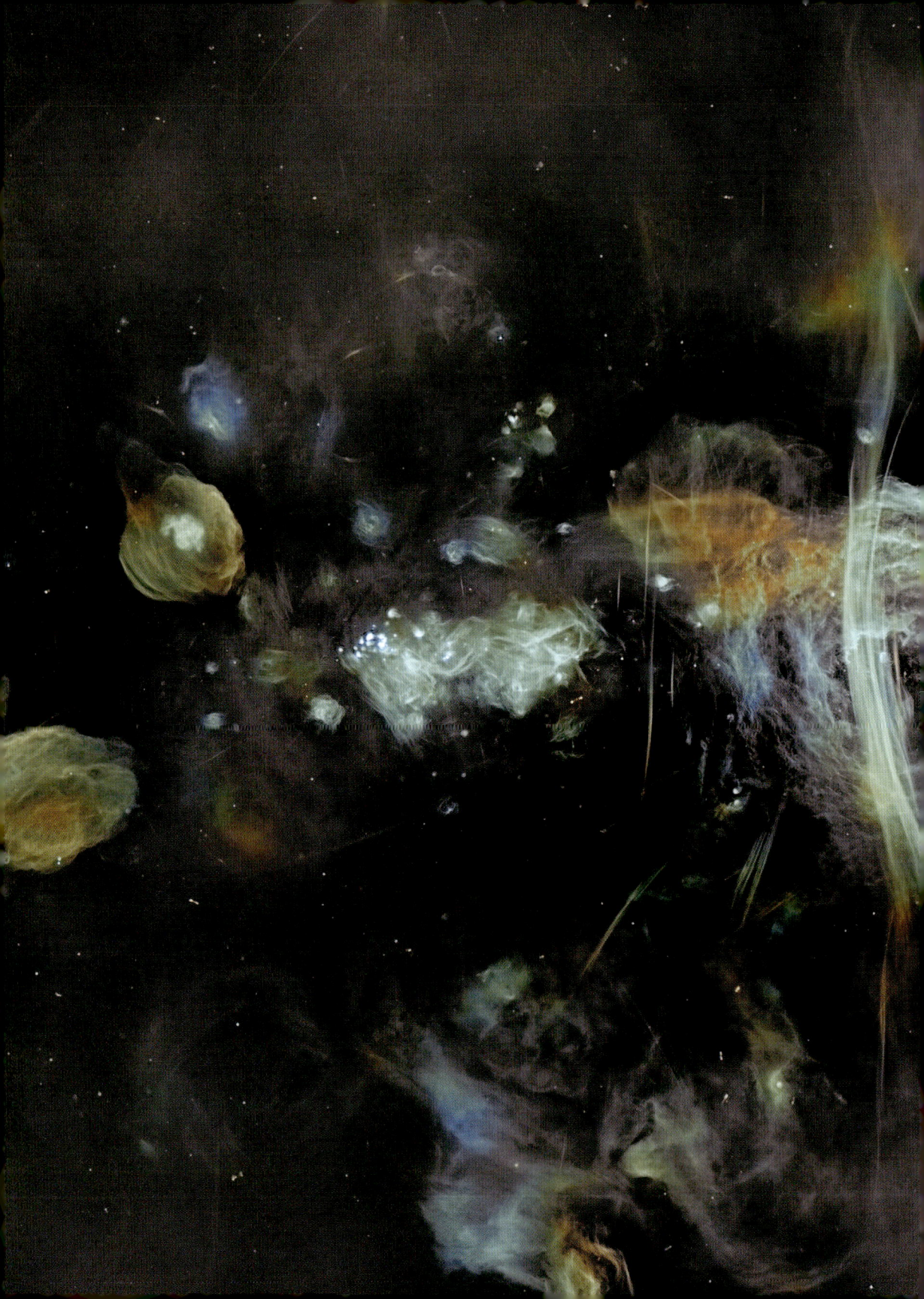

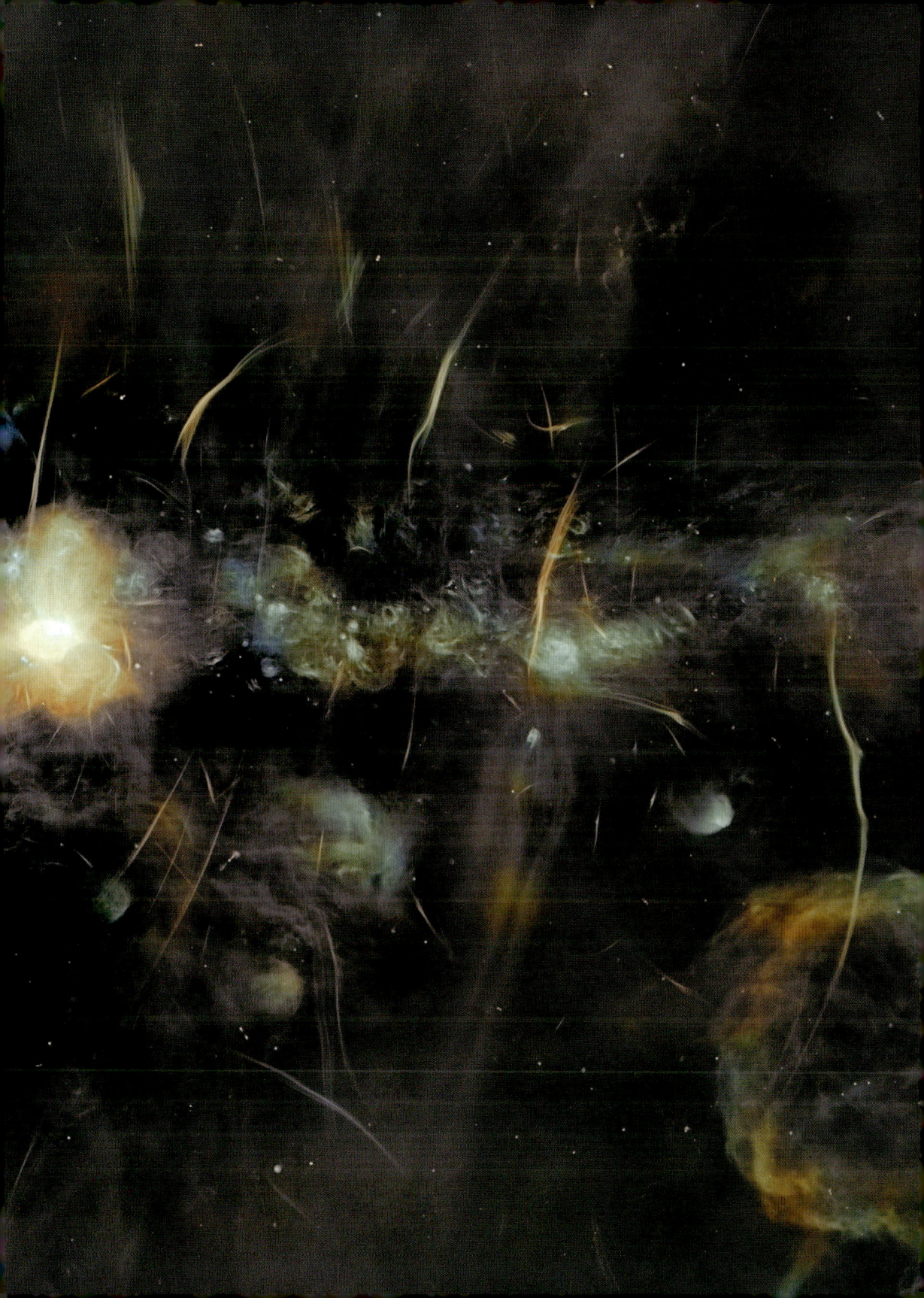

4 million times the mass of the Sun. In other galaxies, supermassive black holes may contain *billions* of solar masses. Astronomers assigned the clever name "*ultra*massive black hole" to the 66 billion solar-mass beast known as TON 618, which has more than 16,000 times the mass of Sgr A*. Astronomers wonder if even bigger black holes exist and what sets the upper limit to their size.

Just like stellar-mass black holes, supermassive black holes are ultimately very simple in that they can be described by their size, spin, and charge. But their origins may not be so simple. Supermassive black holes could not have formed from the collapse of a single star—there's no evidence that stars big enough exist. They must have acquired their mass in a completely different way. Adding to the mystery is the fact that we see supermassive black holes in galaxies throughout the cosmos. This means that they started forming very early in the universe's history. How? One possibility is that they formed by the rapid accumulation of stars and smaller black holes, though some research suggests this would take too long. Another idea is that supermassive black holes formed from the direct collapse of gigantic primordial gas clouds in the early universe. According to some theorists, this process may have taken a little more than 100 million years, a twinkling of an eye on cosmic timescales.

Astronomers learn about supermassive black holes by observing how they impact their environment, similar to how they observe stellar-mass black holes. The enormous gravitational influence of Sgr A* causes the nearby stars and gas clouds to zip around it at thousands of miles per second. The orbital speeds of these stars and clouds can be measured to infer its mass. This is what professors Andrea Ghez and Reinhard Genzel did, accumulating decades' worth of observations of the stars and clouds moving in its vicinity. In 2020, they were awarded the Nobel Prize for their independent work, which provided decisive evidence that Sgr A* is a supermassive black hole.

The Event Horizon Telescope (EHT) was dreamed up to directly observe the region immediately surrounding supermassive black holes. Given their stupendous masses, supermassive black holes are *tiny* compared to what you might expect. The size of Sgr A* is only 1/10 the distance from Earth to the Sun, well within the orbit of Mercury! (This is why our Solar System, located in a distant outpost of the Galaxy, is in no danger of being gobbled up by

Sgr A*.) Sgr A* is more than 25,000 light-years from us; its apparent size on the sky is equivalent to how a blueberry would look on the Moon! A telescope with exquisite resolving power is needed to see that level of detail. No single telescope is big enough to accomplish such a feat.

To get around this problem, astronomers use a technique called interferometry. An interferometer is a group of telescopes that act as one to produce images with far greater detail than a single telescope can. The resolution of a single telescope—how small a detail it can see—is limited by the size of the mirror or dish. With an interferometer, the resolution depends on the distance between the individual telescopes. The farther apart you set the telescopes—the longer their *baseline*—the higher the resolution you can achieve. Scientists have built a global network of telescopes separated by continents, thus using the entire planet as an interferometer. This technique is called very long baseline interferometry (VLBI). With telescopes located in North and South America, Europe, and the South Pole, VLBI enabled the Event Horizon Telescope in 2019 to observe the supermassive black hole at the center of the M87 galaxy:* the first ever "picture" taken of a black hole. A few years later, a picture of the Milky Way's own Sgr A* followed.

The supermassive black hole at the center of M87 is 6.5 billion solar masses, 1,600 times more massive than the Milky Way's. This means its event horizon is about 1,600 times larger, too. Yet the images of the two black holes look quite similar, because the physics around these phenomena is universal. The EHT observed the black holes at millimeter wavelengths of light, which the human eye cannot see.† The pictures are false-color images that allow us to visualize what is going on. They each show a glowing ring of gas, which has been warped due to the extreme curvature of space caused by the black hole. The brighter knots visible in the ring could come from clumps of gas, but scientists think they might simply be an outcome of how the observations

---

* The specific locations of EHT telescopes are the USA, Mexico, Chile, France, Spain, Greenland, and the South Pole.

† One millimeter is equivalent to 0.13 inch.

were taken and physically unrelated to the system. The darkened circle in each image does not represent the black hole's event horizon but rather the shadow caused by its gravity. The size and shape of the shadow are determined by the spin of the black hole as well as by severe gravitational lensing. Just like a glass lens, gravity can distort the appearance of objects. Scientists used the observations to calculate the sizes of the supermassive black holes.

Not long after the Event Horizon Telescope released its first snapshot of a black hole, a separate team of scientists announced the discovery of a *pair* of supermassive black holes in the galaxy NGC 7727. Located 89 million light-years away, this is the closest known binary supermassive black hole. Like the much smaller stellar-mass binaries detected by LIGO, these supermassive black holes are locked in orbit by their mutual gravity and are destined to collide. But LIGO is too small a facility to detect signals from supermassive black hole collisions. This problem may be solved not too long from now. These more massive mergers produce larger ripples in spacetime, which require observatories with much longer arms to detect them. The Laser Interferometer Space Antenna, LISA, is expected to become the first space-based gravitational wave observatory. It is set to launch in 2037, and the latest reports say it will have arms *1.6 million miles (2.5 million kilometers)* in length, roughly 200 times the diameter of Earth! LISA will be able to directly detect the most massive kinds of black holes in the universe, as well as smaller ones in the Milky Way that escape detection by LIGO.

Scientists estimate that once every month or so, within a billion-light-year radius of Earth, at least one pair of black holes comes together in a final embrace. Like a mountain dropped into the cosmic ocean of spacetime, they launch enormous waves that steadily subside into gentle ripples by the time they reach our terrestrial shore. To contemplate their dark dance is to be reminded, as Jonathan Swift said, that "vision is the art of seeing what is invisible to others."

Touching the Invisible

Both scientist and artist delight in making the invisible visible. The scientist wants to excavate the secrets of nature. The artist feels compelled to shed

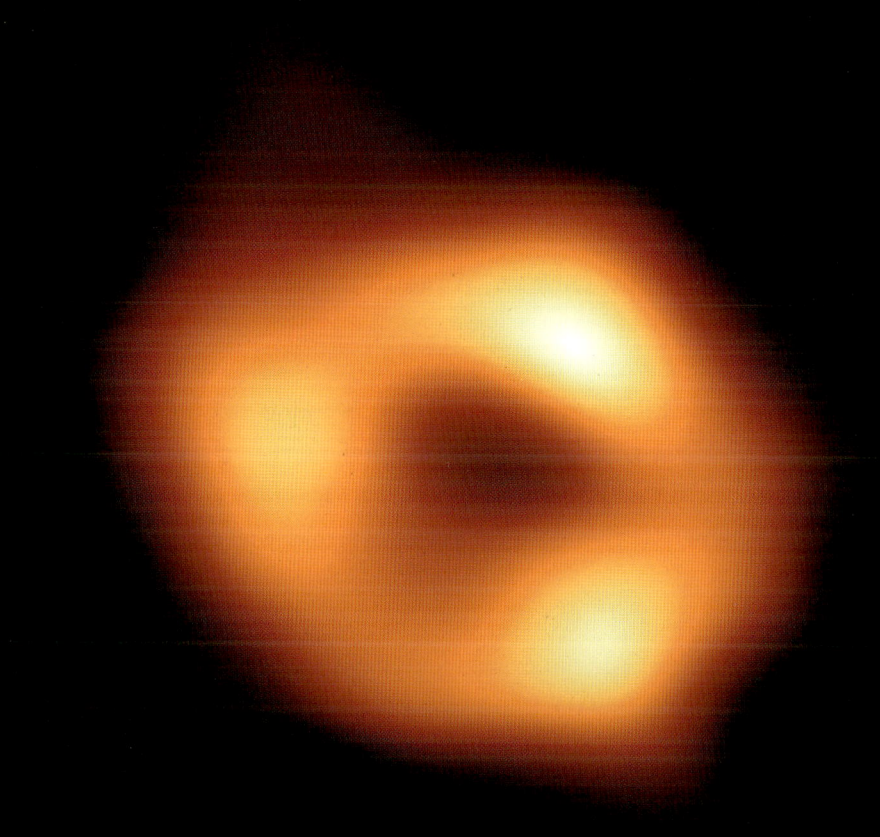

SAGITTARIUS A* | Picture of the Milky Way Galaxy's supermassive black hole, Sagittarius A*, taken by the Event Horizon Telescope. *CREDIT: EHT COLLABORATION.*

light on the secrets of the human personality. Mae Jemison put it this way: "Science provides an understanding of a universal experience. Arts provide a universal understanding of personal experience." When these two intentions come together, they can give birth to magical moments—as when we realize for the first time that there are places in the universe where light can't escape or that we're all connected by the fabric of spacetime. "The really wonderful thing that happened to me when I was in space," said Jemison, "was this feeling of belonging to the entire universe." We are *all* in space. All of us are children of the cosmos.

In 1992, the year that Mae Jemison became the first black woman to travel to space, the artist Elizabeth Catlett was busy working away in her studio in Cuernavaca, Mexico. She was 77 years old, and for most of her long career, her leading objective as an artist was to create art for ordinary people that affirmed and honored the dignity, grace, power, struggles, and beauty of black women.

One of the pieces she completed that year is a portrait of a woman whose stylized face suggests a West African mask, a motif she developed early on and returned to often. The folds of the woman's dress, which include collaged fabric, appear to be inspired by cubism, an art movement that heavily appropriated ideas and images from West African art. The woman is depicted from the torso up, emerging from a platform as though she is a sculpture, another of Catlett's favored art forms. The portrait is called *Harlem Woman*, inspired by a lady the artist saw standing one day on 125th Street. In connecting multiple art traditions, Catlett links black American women to the rich cultural and spiritual heritage of their foremothers. With images like *Harlem Woman* and *Madonna*, completed a decade earlier, Catlett situates black women in a broad global and historical context, raising them out of the isolation and marginalization so characteristic of their American experience.

Speaking of the relationship of technique to the deeper intention and meaning in her work, Catlett said:

I still work figuratively, trying to express emotion through abstract form, color, line, and space. I attempt to reach out to ordinary people

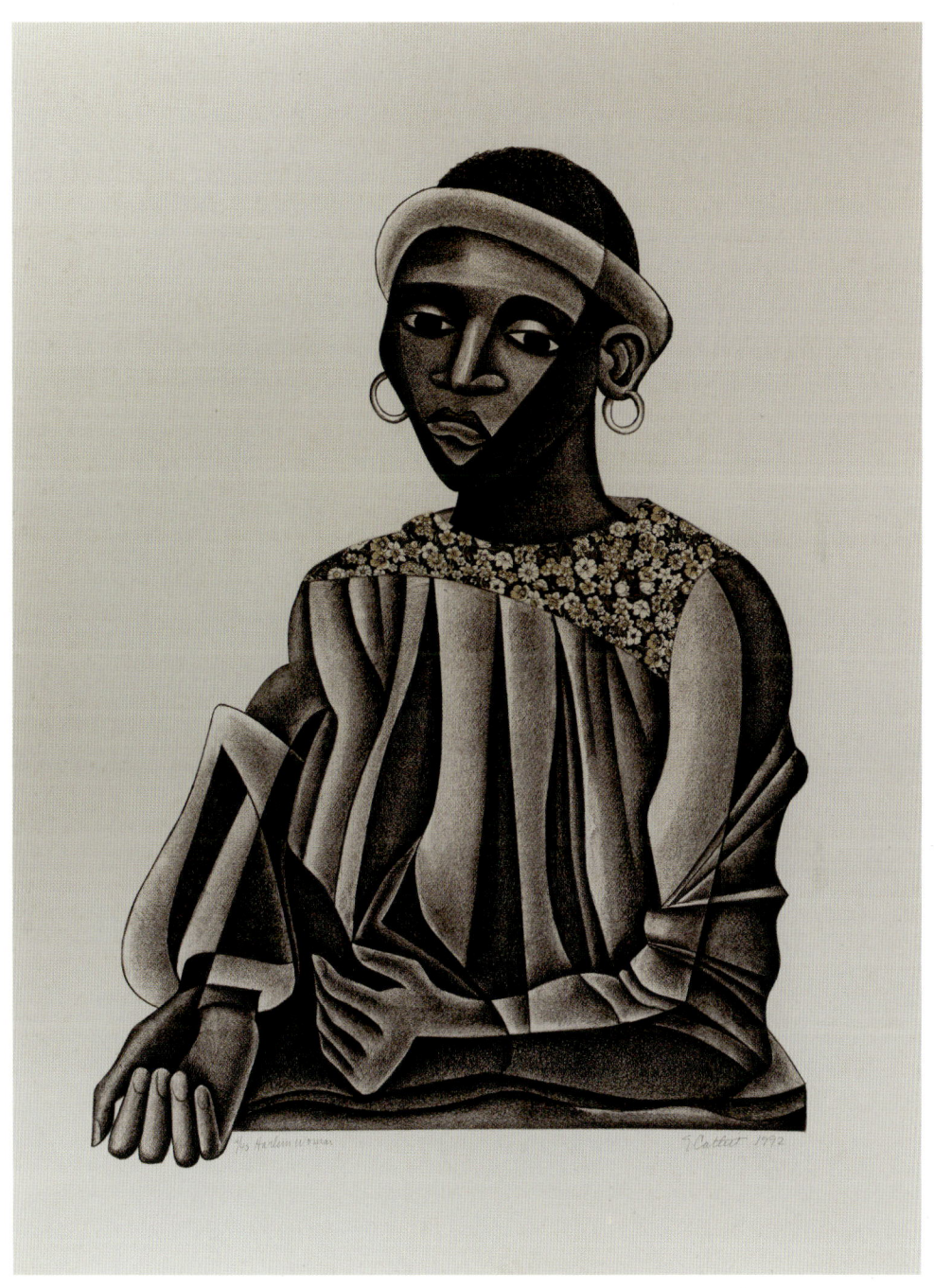

HARLEM WOMAN | Elizabeth Catlett, 1992, lithograph with fabric collage, 24 × 17⅛ in (61 × 43.5 cm). © 2024 Mora-Catlett Family / Licensed by VAGA at Artists Rights Society (ARS), New York.

who have little or no experience or understanding of art principles, and extend to them what I may feel about a subject, whether anger, indignation, strength, beauty, sacrifice, understanding—whatever, but always something. Even in more abstract sculpture I attempt to get a reaction, possibly by a strong upward gesture, or a close tight feeling between two figures.

Catlett believed that artists have a responsibility to illuminate injustice and reveal the truth about people who suffer abuse at the hands of the powerful. As a black woman who found greater opportunities and freedom as an artist in Mexico than she did in the United States, she understood that change was not only necessary but possible. "Art cannot effect this change," she explained, "but it can raise the consciousness of the necessities and the possibilities of doing so."

The Negro Woman series of 1946–47 exemplifies Catlett's efforts to remove black women from invisibility, her powerful commitment to transforming the way they are typically treated in the culture—as objects—by explicitly representing them as who they are: autonomous subjects. The series includes images of both ordinary unnamed women and famous historical figures, including Phillis Wheatley, Sojourner Truth, and Harriet Tubman. Each linocut in the series has a title containing an "I" or "my" statement that forces the viewer to identify with the subject of the image:

"I am the Negro woman"—"I have always worked hard in America"—"I have given the world my songs"—"In Harriet Tubman I helped hundreds to freedom"—"In Phillis Wheatley I proved intellectual equality in the midst of slavery"—"My right is a future of equality with other Americans."

By placing domestic workers, blues singers, sharecroppers, and other everyday women side by side with celebrated leaders, Catlett reminds viewers not only of black women's essential role in modern society but of their historical contributions to the advancement of humanity. In a Western art tradition that typically presents black women and other women of color as objects, *The Negro Woman* points a new way forward by reclaiming their rightful place in the historical record. With Catlett's art as a whole, she refuses to let women

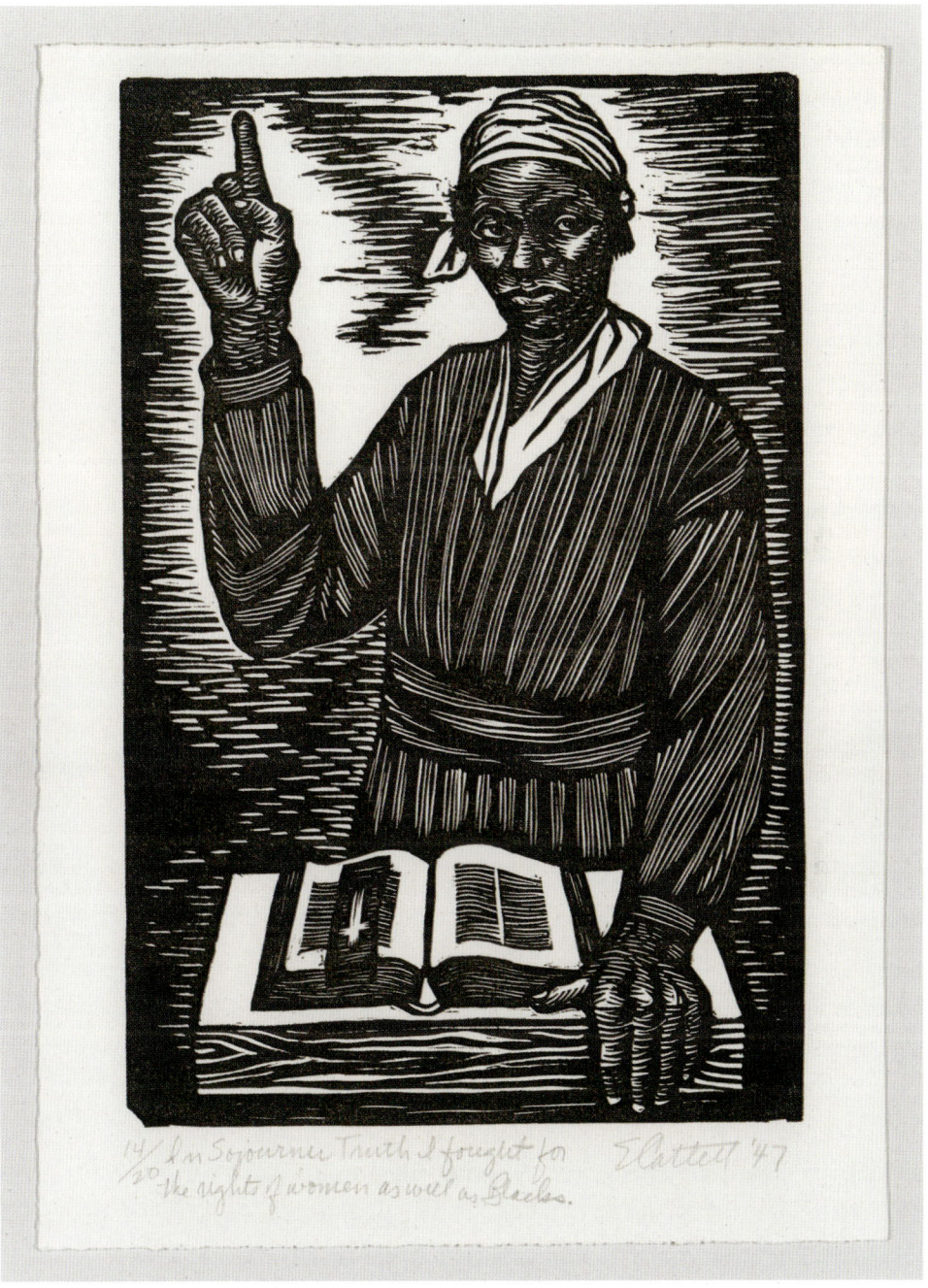

IN SOJOURNER TRUTH I FOUGHT FOR THE RIGHTS OF WOMEN AS WELL AS NEGROES |
Elizabeth Catlett, 1947, printed 1989, linoleum cut, image: 8⅞ × 5¹⁵⁄₁₆ in (22.5 × 15.1 cm).

disappear from history. As she once expressed it, "Women will have to look back at the history of women in art in order to base the future on that history."

In her paintings and sculptures, Catlett's powerful message is heightened through her technical mastery over the use of space. In visual art, *space* refers to the area around, between, and occupied by the various components of a composition. In painting, the area occupied by the main subject (or subjects) of the work is called *positive space*, while the surrounding area is called *negative space*. In sculpture, positive space refers to the physical object created by the artist, and negative space to the area around and between. An exploration of positive and negative space can reveal how an artist employs her medium's technical aspects to express her intentions and to evoke an emotional response in the viewer. An artist's use of space is an important element of her unique style and language.

Negative space may be used to emphasize the subject of the artwork, to imply the artist's focus and intent. But some artists, such as Romare Bearden, don't buy into the idea of two distinct spaces. "There is really no such thing as a 'negative space,'" he wrote in his journal. "In a well-organized picture all the spaces must be considered as being positive functions towards the ordering of the whole." Kara Walker, who has been so successful at blurring the boundaries between positive and negative space in her striking silhouettes, might agree.

In her *Negro Woman* series, Catlett approaches space in multiple ways. In each linocut, she fills almost the entire frame with the face or body of the subject, leaving very little negative space. For instance, in *Sojourner Truth*, the heroine's figure occupies nearly the whole picture, her head and raised right hand grazing the very top of the image, her left hand touching the bottom edge. This creates a sense of intimacy between her and the viewer.

Secondly, Catlett flattens the images, so that the foreground subject and background appear projected onto one plane. The images lack a "realistic" sense of perspective. Again, this implies closeness, this time between the subject and the objects, people, or ideas that are at the center of her attention. A similar effect is achieved by Hokusai, one of the great Japanese artists of the Edo period. In paintings like *The Waterfall Where Yoshitsune Washed His*

Horse at Yoshino in Yamato Province, the closeness and immediacy of nature are powerfully evoked by projecting everything onto the same surface. In Catlett's series, *Harriet Tubman* points the way North with a strong, outstretched arm that extends directly over the heads of the escapees, like a channel for the flow of energy that propels them forward to freedom. In *Special Houses*, two women are thrust to the front of the image by a row of stark tenement houses, a literal and allegorical representation of the unjust, miserable living conditions they endure every day.

Another way artists approach space is through the relative sizes and positions of objects. In Catlett's *Phillis Wheatley*, this technique is used in combination with her characteristic flattening of images. Here the poet contemplates, quill pen in hand, while symbolic representations of her thoughts—enslaved women like her—stand next to her head.

In traditional landscape painting, the relative sizes of objects indicate their relative spatial positions. Smaller objects are perceived as literally farther away. In images like *Phillis Wheatley* and *Blues*, relative size evokes an *emotional* and *metaphorical* sense of space, rather than a literal one. Captured by slave traders from her home in Gambia at the age of seven or eight, Wheatley was the first black woman to publish a book of poetry in America. These words from her poem "To S.M., a Young African Painter, on Seeing His Works" could be projected nearly two centuries into the future to describe the impact of viewing Catlett's portrait of her:

> To show the lab'ring bosom's deep intent,
> And thought in living characters to paint,
> When first thy pencil did those beauties give,
> And breathing figures learnt from thee to live,
> How did those prospects give my soul delight
> A new creation rushing on my sight?

Catlett's approach to space in sculpture in many ways parallels her approach in her two-dimensional work. Her numerous *Mother and Child* sculptures in terracotta, mahogany, and bronze are sturdy, tight pieces without much negative space running through them; their physical form mirrors

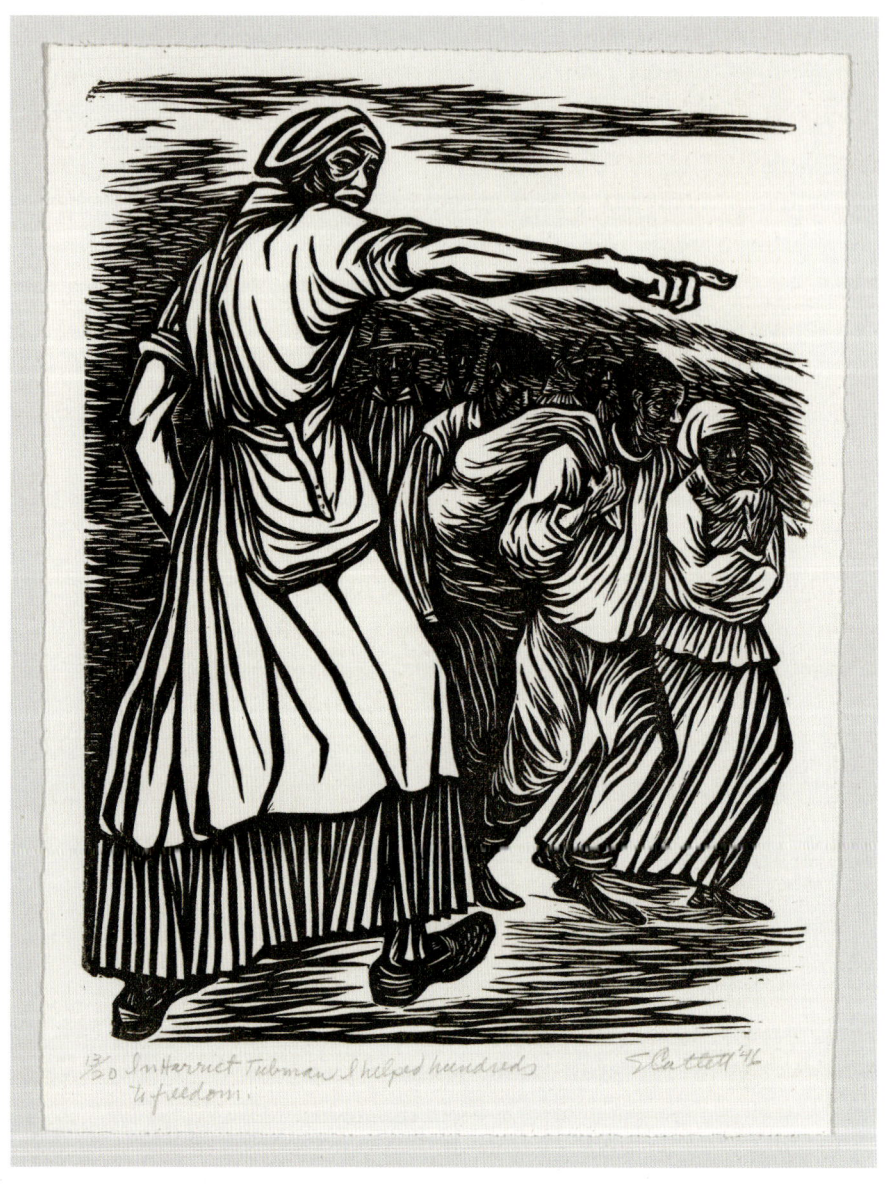

IN HARRIET TUBMAN I HELPED HUNDREDS TO FREEDOM | Elizabeth Catlett, 1946, linoleum cut, image: 9⅛ × 7¹/₁₆ in (23.2 × 17.9 cm). © 2024 *Mora-Catlett Family / Licensed by VAGA at Artists Rights Society (ARS), New York.*

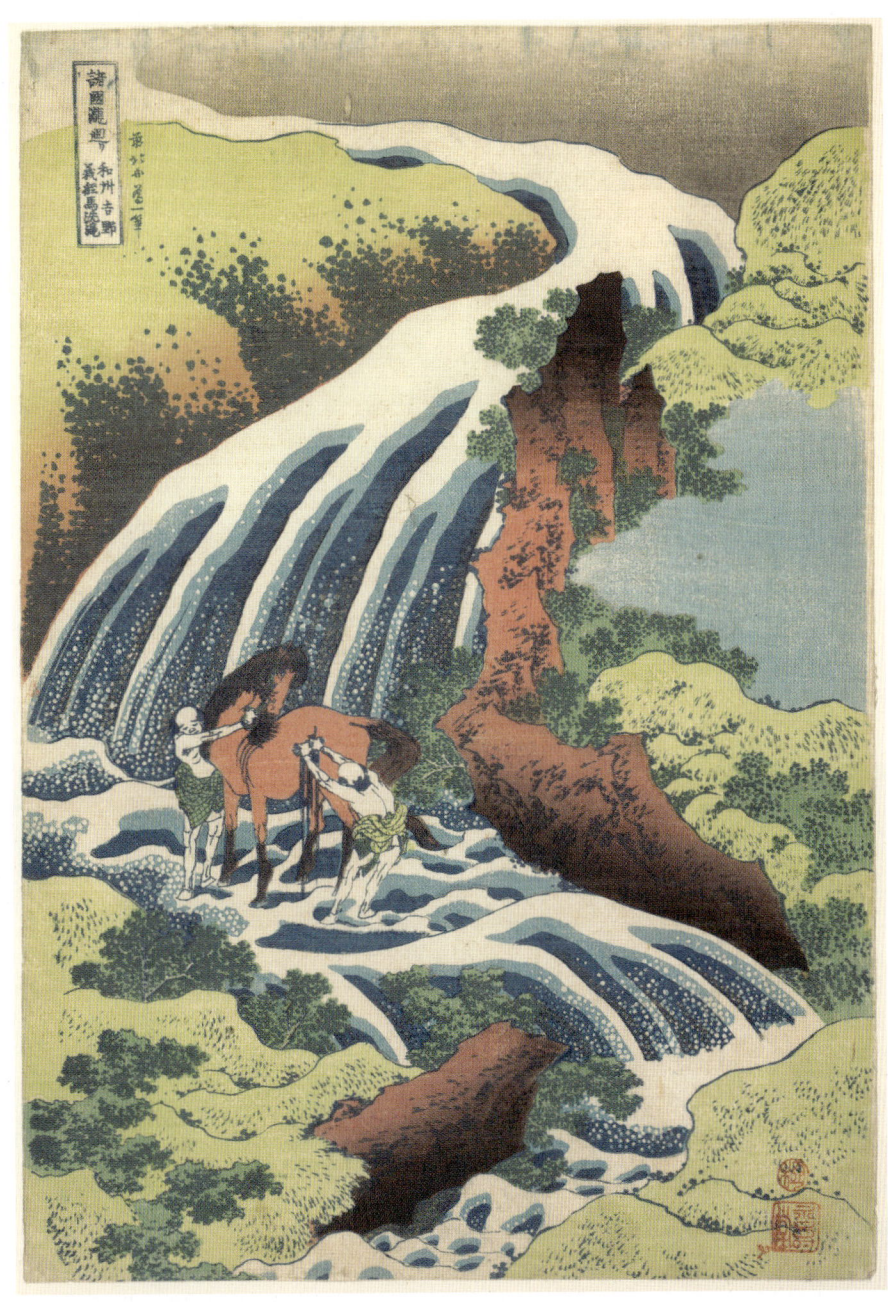

THE WATERFALL WHERE YOSHITSUNE WASHED HIS HORSE AT YOSHINO IN YAMATO PROVINCE | Katsushika Hokusai, ca. 1832, woodblock print; ink and color; vertical ōban, 14⅝ × 10¼ in (37.1 × 26 cm).

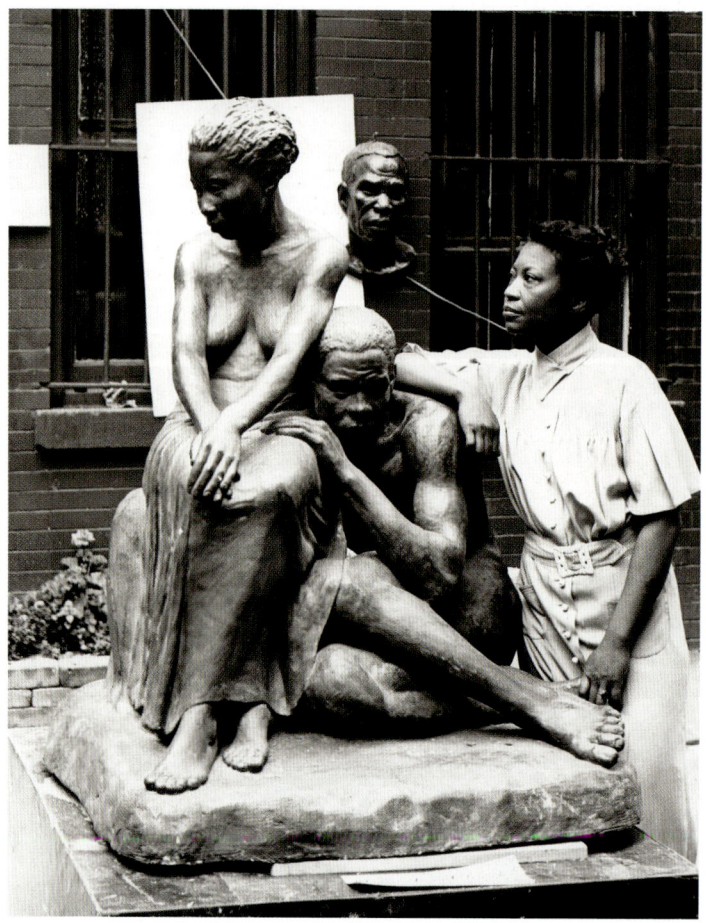

AUGUSTA SAVAGE | The artist leaning on her sculpture *Realization*, around 1938.

the intimacy and solidity of the mother-child bond. Catlett's identity as a woman and mother was integral to her art, and it fed her commitment to making art that ordinary women and men could relate to. So much so that she would act out the poses of her sculptures in order to fully explore her feeling. "It's feeling more visual when I'm working out an idea," she said. "So let's say it's a model sitting, leaning on an elbow. I'll sit like that, and I'll feel where the stress and tensions are."

Contrast the density of some of Catlett's sculptures to the airiness of works by Augusta Savage, such as *Pumbaa* or *The Diving Boy*, which are much

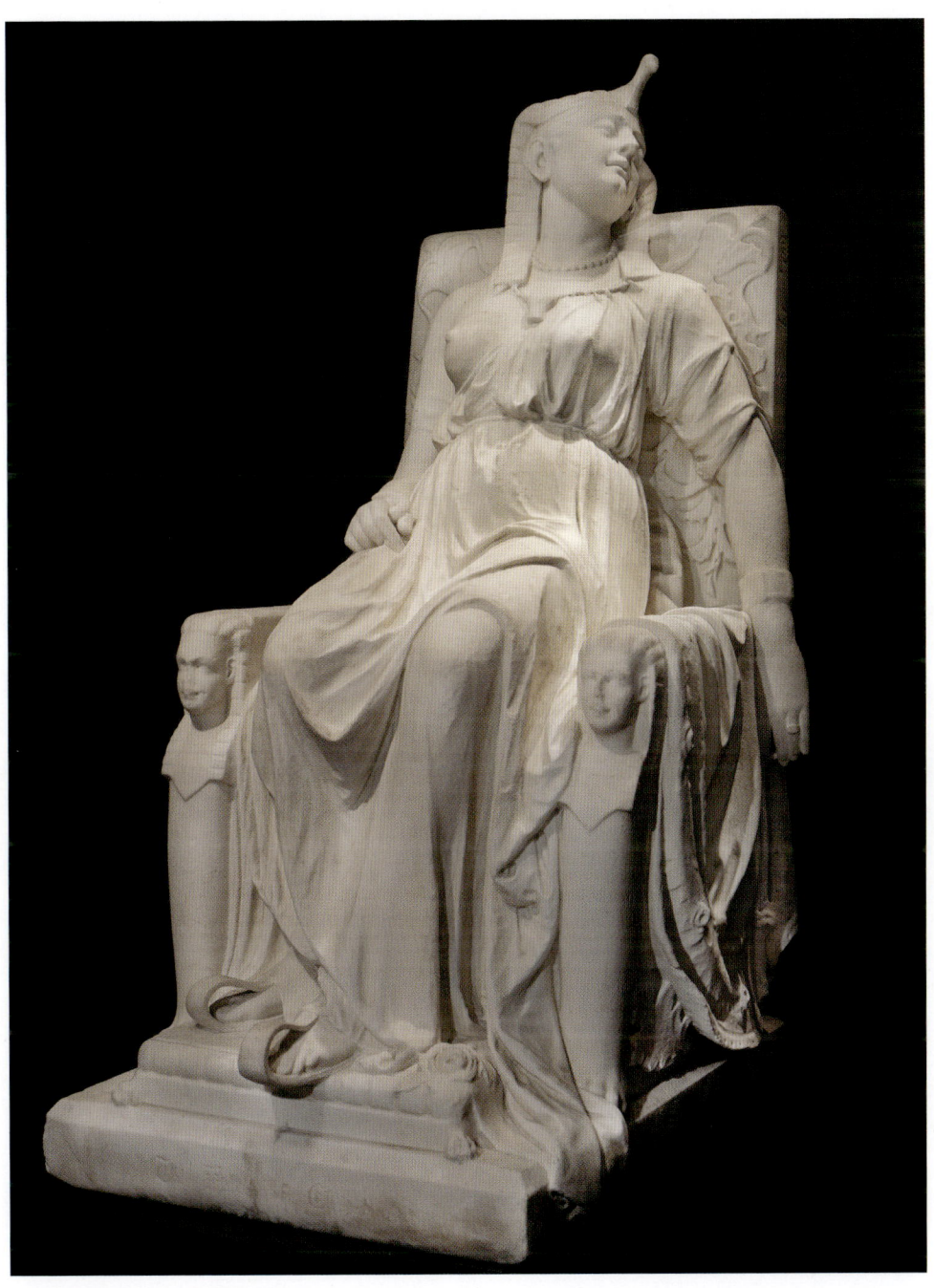

THE DEATH OF CLEOPATRA | Edmonia Lewis, 1876, marble, 63 × 31¼ × 46 in (160 × 79.4 × 116.8 cm).

more liberal in their use of positive space. Or compare the smooth, simple contours of one of Catlett's standing woman sculptures to the intricate, undulating curves with which Meta Vaux Warrick Fuller transforms space in *Danse Macabre*. And then there's Edmonia Lewis, whose monumental works of carved marble depict the human form—including epic figures like Cleopatra, Hagar, and Tecumseh—in sweeping, sensual, hyperrealistic gesture.

Born in Washington, DC, in 1915 and an émigré to Mexico, Catlett was part of a community that extended across both space and time. She belonged to a lineage of black women sculptors who boldly inserted their new vision into the canon of Western art. She followed Lewis (1844–1907), Fuller (1877–1968), and Savage (1892–1962), artists who transformed space in their own distinctive ways to express ideas that converged more than diverged. Lewis was born during slavery and, like Catlett, was compelled to leave America to find a modicum of freedom and respect as an artist. Both Fuller and Savage flourished in the artistically fertile soil of the Harlem Renaissance, where they drew inspiration from and found community with other prominent artists and intellectuals of the time.

Each of these women mastered the ability to give tangible expression to abstract ideas and revolutionized how we think about identity, womanhood, the black experience, and the human condition. Like ripples of space set in motion by distant black holes billions of light-years away, their art enabled people to make contact with a world that, if it were up to some, would remain invisible.

CHAPTER SEVEN

TIME—LIGHT'S MEMORY

The only reason for time is so that everything doesn't happen at once.
— Albert Einstein —

From the experience of the past, the present acts prudently,
lest it spoil future actions.
— From the inscription on Titian's painting
Allegory of Time Governed by Prudence —

It is looking at things for a long time that ripens you and gives you
a deeper meaning.
— Vincent van Gogh —

T ime is the measure of change. If the seasons didn't turn, children didn't grow older, and the universe didn't evolve, there would be no need for the idea of time. But everything changes, and we all must obey time's unbending law. In the West, time is perceived as linear, an arrow that points in only one direction. Many indigenous cultures in Asia, Africa, and the Americas view time as cyclical; time is repetitive and continuously returns upon itself. Given the diverse conceptualizations of time across different cultures, it would be tempting to think that it's simply a human construct. But in fact, time has an existence independent of us. It's a fundamental part of the fabric of reality, and none of us can evade its laws any more than we can defy gravity or stop the Sun from shining.

The physicist thinks of time as one of the four indispensable numbers used to describe an event that takes place at any moment in the history of the universe. Three numbers are needed to define where the event took place, and the fourth, what time it occurred. So we can say that Vincent van Gogh cut off a bit of his ear at a house in Arles—a town located some 460 miles (740 kilometers) south and 75 miles (120 kilometers) east of Paris, at about sea level*—on the night of December 23, 1888. On a cosmic scale, I can tell you that the Crab Nebula is located at a Galactic longitude of 184 degrees, a latitude of minus 5.7 degrees, at a distance of 6,500 light-years from Earth. When did the star that would become the nebula explode as a supernova? It was first noticed by Chinese astronomers and others around the world in 1054, as was mentioned earlier. Since the light from the supernova was already 6,500 years old by the time they saw it, that means the actual explosion took place about 7,554 years ago.

This way of thinking is intuitive to most of us. We recognize that both time and place are essential pieces of information required to locate events in the past and to prepare for the future. When planning for a wedding, the bride and groom must agree upon both the where *and* the when. The guests can move through space to get to the wedding. But anyone who misses the big day cannot, as far as we know, travel back through time to attend. The universe gives us a limited freedom—we can move freely through space, but we have less liberty in how we move through time.

---

* More precisely, Arles has a longitude, latitude, and elevation of 4.6 degrees, 43.6 degrees, and 1 yard (about 1 meter), respectively.

We also recognize intuitively that time and motion are linked. To say that someone is 30 years old means that the Earth has moved through space around the Sun 30 times since that person's birth. Motion, time, and space are, in fact, physically related. From the time we were kids playing racing games, we've been aware that the time it takes to cover a distance depends on how fast you're going. What would happen if you could run *really* fast? Like, light-speed fast? Albert Einstein's childhood games were a bit different from many of ours. At the age of 16, he imagined what it would be like to chase a beam of light. He was curious about the motion of light, and this curiosity gave rise to insights that revolutionized how we think about time and space.

A Persistent Illusion

Einstein knew about the work of the Scottish physicist James Clerk Maxwell, who showed the world that light, magnetism, and electricity are all different aspects of the same phenomenon. Maxwell was the first to recognize that the speed of light is always constant in a vacuum. This key fact was the cornerstone of a famous thought experiment that led Einstein to some incredibly counterintuitive conclusions about the nature of reality.

Imagine an observer named Paul standing on a railway platform as a train goes hurtling by. On either side of him, at equal distances, are two trees. All of a sudden, just as the midpoint of the train is passing Paul, the trees are struck by bolts of lightning. Since the two lightning bolts must travel the same distance to Paul, their light reaches his eyes at the same moment. He sees the trees being hit at the same time.

Now imagine a second observer, Vincent, sitting in the middle of the moving train. From his position, the light from the tree that the train is approaching reaches him first, since it travels a shorter distance. Vincent concludes that this tree is hit first. He and Paul reach different conclusions as to whether the lightning strikes happened simultaneously, and both of them are correct.

Based on your everyday experience, your common sense might have guessed (wrongly) that both Paul and Vincent saw the event happening at the same time, since they were both situated halfway between the light sources.

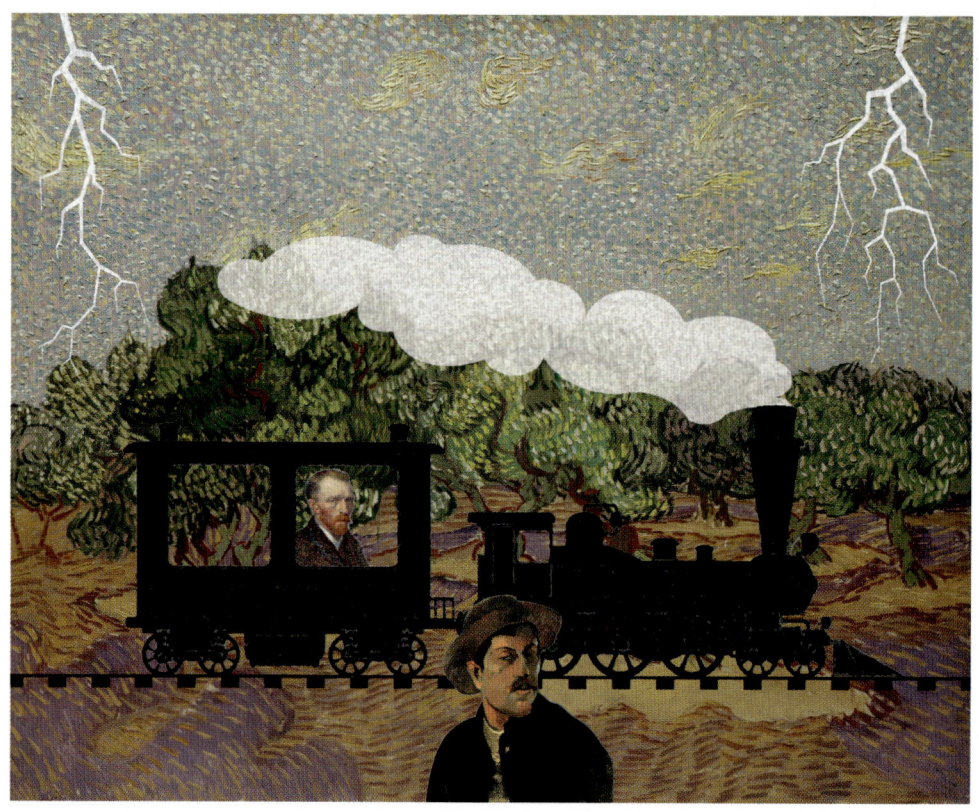

MY IMAGINING OF EINSTEIN'S THOUGHT EXPERIMENT ON SIMULTANEITY | As the train passes by Paul, standing outside, two lightning bolts strike on either side of the train at equal distances. Paul sees the bolts strike at the same time. But Vincent, inside the train, sees the lightning bolt that the train is approaching first.

But Einstein showed that because the speed of light is constant and measured to be the same by all observers, this cannot be the case. Whether events appear to happen simultaneously is *relative.*

This has a deep implication for the nature of time. The fact that an event can appear to happen simultaneously for one observer and not for another means that *time flows differently for them*, depending on their state of motion. For Vincent, in the moving train, it turns out that time slows down.

Einstein realized, moreover, that if the speed of light is measured to be the same by all observers, then old concepts about fixed time and space would have to be exchanged for new ones that depend on the observer's state of motion. A key idea of Einstein's theory of special relativity is that time and space are part of the same indivisible "fabric" of the universe. They make up an interdependent, four-dimensional *spacetime*, consisting of the three dimensions of space and time's single dimension. The four-dimensional universe is *everything* there is—not only all of space, but everything as it exists at each moment in time—past, present, and future.

In most of our day-to-day ordinary activities, relativistic effects are imperceptible. Compared to light, trains and other vehicles are slow moving; in most cases, our commonsense understanding of nature works just fine. If Paul and Vincent were both stationary observers or moving only slowly with respect to each other, they would agree on when and where an event took place. But funny things happen when you begin to approach the speed of light.

Three things occur: time slows, space shrinks, and mass increases. Imagine that now you are the pilot of a futuristic spaceship traveling between two locations at half the speed of light. Observers on Earth measure the distance between the locations to be a certain number. But you and your fellow astronauts measure a *shorter* distance. For you, the pilot, space literally contracts. Since space and time are interdependent, any change in space (that is, distance or length) produces a change in time, and vice versa. The laws of nature—the relationship between time, distance, and speed—work the same both on and off the spaceship. So not only does space get shorter for those of you on the spaceship, but also time dilates. The clock on your ship actually ticks *slower* than the clocks back on Earth, *but for you, time appears to be moving at its regular pace.* When you return home, you will have *aged less* than your Earthbound friends!

Now suppose that you fire up the engine of the spaceship and try to go faster. What happens? According to special relativity, as you accelerate, your mass (as well as the mass of the spaceship and everything in it) goes up. The faster you try to move, the more your mass increases, and the more energy it takes to power the ship. Meanwhile, space contracts further and further, and time slows down more and more. It would take an infinite amount of power to reach the speed of light. An impossible feat. Nothing in the universe can overshoot this universal speed limit.

It is fun, however, to imagine what might happen if you could . . .

There once was a lady named Bright,
Who traveled much faster than light.
She departed one day,
In a relative way,
And returned on the previous night.*

Remembering Time

We are all time travelers, moving along with the universe into the unknown at the obstinate rate of one second per second. We can't change the past; in that sense it is locked to us. Yet we can access it through books and other media, art, and culture—and our own memories—and exercise our power to decide its meaning. Only the immediate future can be anticipated with any level of confidence, and that, just barely. Nevertheless, the past and future are as real as the present. In some sense we are time's captives—obediently organizing our lives around clocks and calendars that we played no part in establishing. Yet in another very real sense, the way we experience life depends on how we view and experience time. After learning about the death of a close

---

* Some people attribute this limerick, published in its original version in 1923, to botanist A. H. Reginald Buller.

friend, Einstein wrote to the family, "Now he has departed from this strange world a little ahead of me. That means nothing. People like us, who believe in physics, know that the distinction between past, present, and future is only a stubbornly persistent illusion."

At the other end of the science-art spectrum, the French novelist Marcel Proust echoes the teachings of special relativity, writing, "The time at our disposal each day is elastic; the passions we *feel* dilate it, those that inspire us shrink it, and habit fills it." The inner, personal world of emotions parallels the external reality of the material world. In both the relativistic and Proustian sense, the experience of time is subjective.

Science has done well at explaining the physical nature of time and describing how the laws of nature link past, present, and future; but conceptually, time remains a "stubbornly persistent illusion" for even the most advanced minds. As familiar as time is, compared to space, it feels more elusive and puzzles the mind with its mysteries. Why can't we move through time as freely as we move through space? Whether it moves in circles or a straight line, what is the origin of time? Is time finite or infinite? What is eternity?

Art comes to our aid in making the intangible concrete. Art is fundamentally allegorical. It makes use of symbols to express abstract ideas and deeper meanings. The abstract idea might be time, love, nationality, or art itself. The symbol used to represent the idea could be color, musical notes, a person, thing, or story. In Titian's famous painting *Allegory of Time Governed by Prudence*, the artist portrays three human heads representing youth, adulthood, and old age, thereby communicating the passage of time in one static image. The heads look in different directions—as though toward the past, present, and future—a visual metaphor that underscores the overall symbolic meaning of the painting. Gustav Klimt's *The Three Ages of Woman* uses a similar metaphor; though with the women's nakedness on display and their striking physical postures, the painting hits with a very different emotional impact.

In art, the time and place of creation are significant because those four numbers provide historical and cultural context. A work of art is permeated with the personality of the artist and the culture in which it is created. When

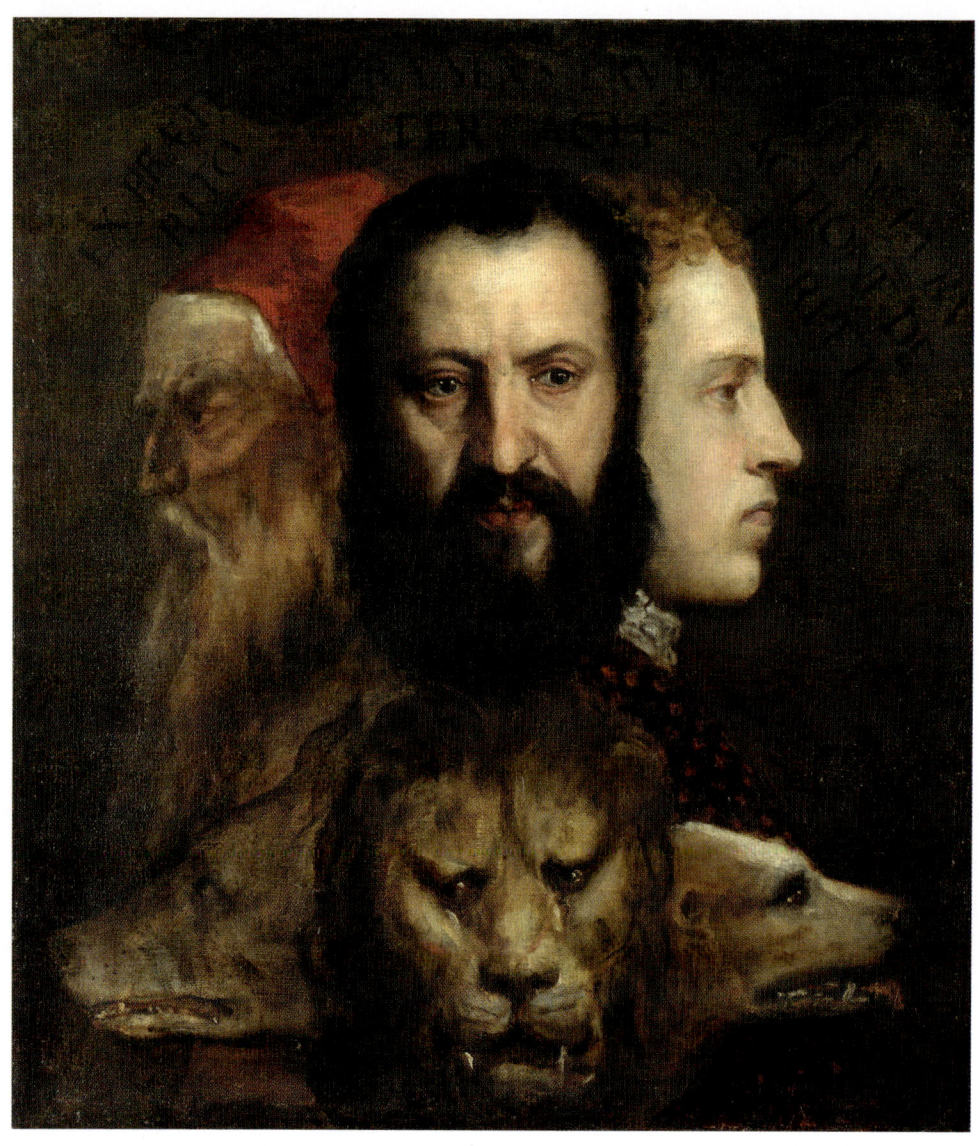

ALLEGORY OF TIME GOVERNED BY PRUDENCE | Titian, ca. 1550, oil on canvas, 29⁷⁄₁₀ × 26⁹⁄₁₀ in (75.5 × 68.4 cm).

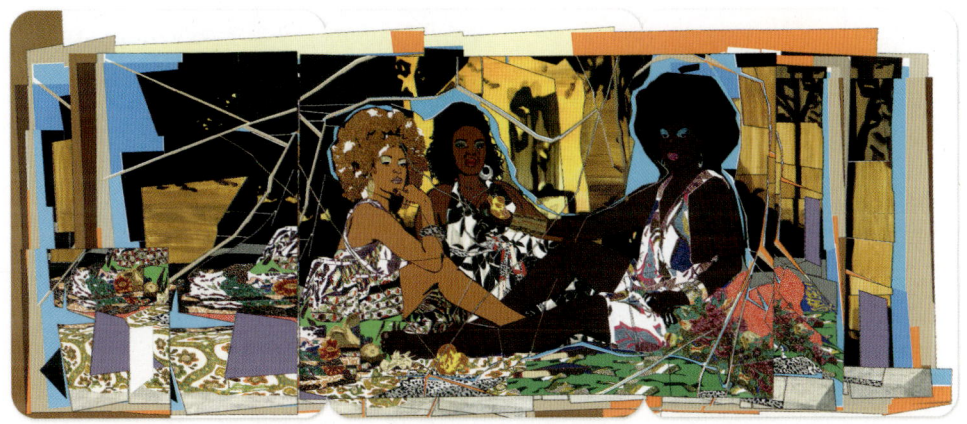

LE DÉJEUNER SUR L'HERBE: LES TROIS FEMMES NOIR | Mickalene Thomas, 2010, rhinestone, acrylic, and enamel on panel, 120 × 288 in (304.8 × 731.5 cm).

the artist creates, she is in dialogue not only with her contemporaries but with artists of the past and future.

The work of the contemporary artist Mickalene Thomas allows us to eavesdrop on a fascinating conversation. Her 2014 painting *Le Déjeuner sur l'herbe: Les Trois Femmes Noir* makes direct reference to Édouard Manet's *Le Déjeuner sur l'herbe*, painted more than a century before.* Picasso also created some 40 variations of this classic of the Western canon. Manet's painting, in turn, was inspired by *Le Concert Champêtre (The Pastoral Concert)* by Titian, as well as another definitive work of the early 16th century, *The Judgment of Paris,* an engraving designed by Raphael.

Like Manet before her, Thomas takes an old motif and, breaking with past tradition, makes it her own. Instead of naked white men or women lounging on the grass, Thomas centers three black women decked out in vibrant 1970s fashion, all three of them gazing with quiet confidence at the viewer. Now the women are in an interior setting; and the large multimedia canvas incorporates textiles, wood panels, and rhinestones, further conjuring the decade of Thomas's childhood. In pointing back through several centuries

---

* In English: *The Luncheon on the Grass: Three Black Women.*

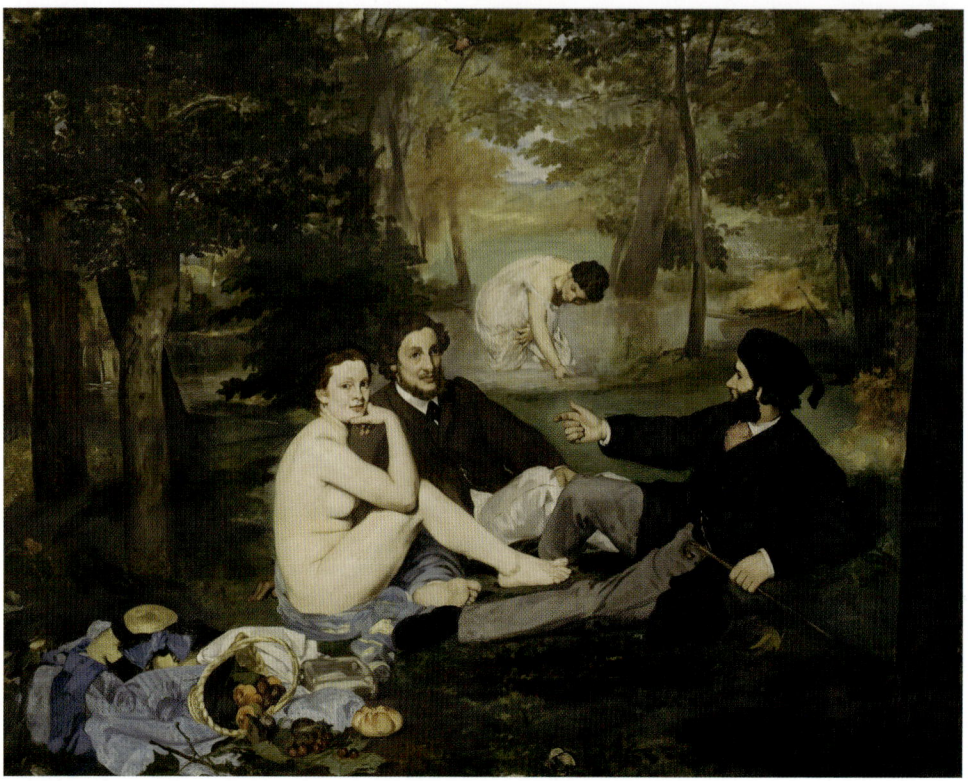

LE DÉJEUNER SUR L'HERBE | Édouard Manet, 1863, oil on canvas, 81⁹/₁₀ × 104¹/₁₀ in (208 × 264.5 cm).

of art history and weaving it with personal history, Thomas asserts that black art and black women are part of the lineage of Western art. What's more, she sets an old idea into an entirely new cultural context and thereby breathes new meaning into old forms.

The artist's work is a creative response to other art; it is the manifestation of the knowledge and impressions that have been accumulated through the filter of personal experience. "To me, if there is any sort of value added to the accumulation of knowledge over time," said Kerry James Marshall, one of the greatest living painters, "then the work of artists should be a reflection of that accumulated value, accumulated knowledge. You have to demonstrate that you have the sophistication to put that into play in the work you're making."

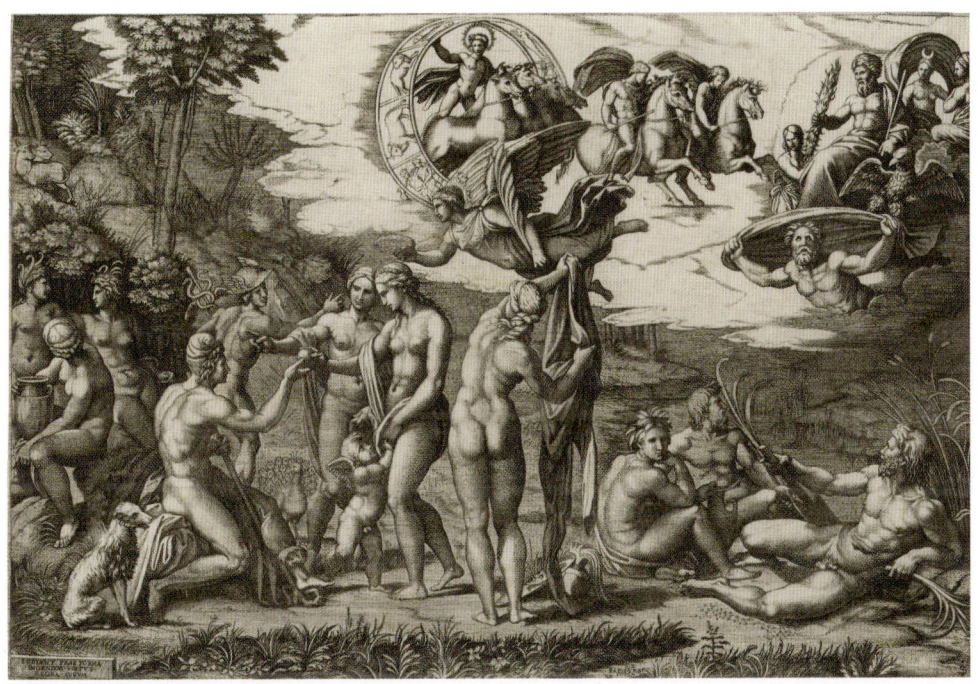

THE JUDGMENT OF PARIS | Marcantonio Raimondi, designed by Raphael, ca. 1510–20, engraving, 11⁷⁄₁₆ × 17³⁄₁₆ in (29.1 × 43.7 cm).

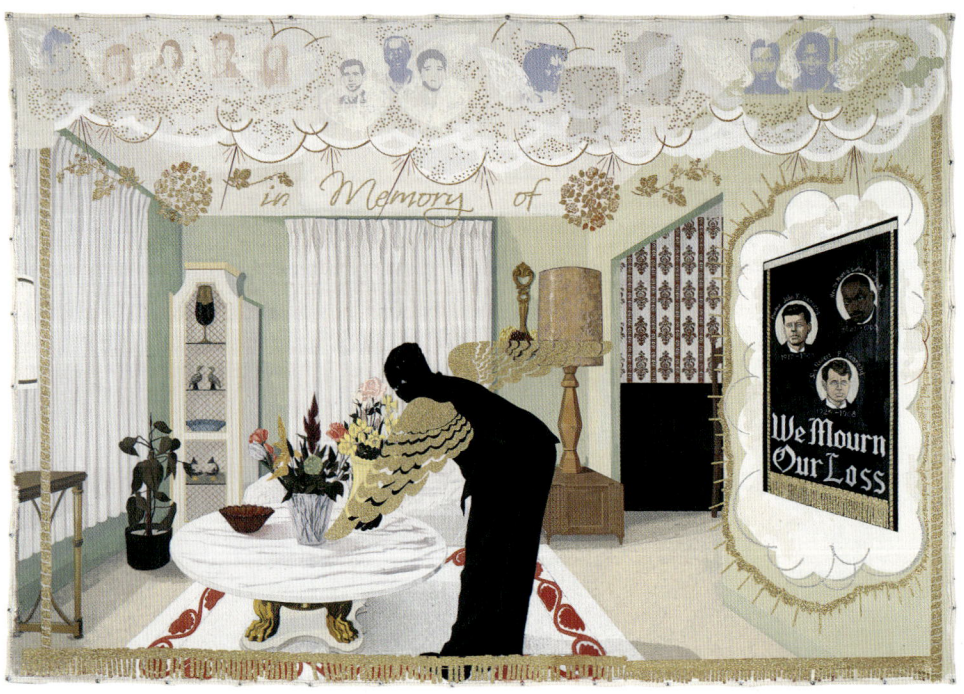

SOUVENIR I | Kerry James Marshall, 1997, acrylic, collage, and glitter on canvas, 108 × 157 in (274.3 × 144.8 cm) © *Kerry James Marshall. Courtesy of the artist and Jack Shainman Gallery, New York.*

In Marshall's *Souvenir* painting series, he commemorates fallen heroes and innocents of the Civil Rights era, as well as prominent artists and intellectuals who transformed American culture. Each of the four paintings is set in what could be your grandma's or auntie's living room—the large, unstretched canvases span 9 by 13 feet, thus lending epic proportions to an otherwise quotidian scene. Marshall's deep knowledge of art history, including art of the Western canon and black art, shines through in the paintings.

In *Souvenir I,* a photo collage of Martin Luther King Jr., JFK, and Robert Kennedy hangs on the wall. This common fixture of many black homes of the era sets the scene squarely in the 1960s. Floating above in clouds, the faces of Medgar Evers, Fred Hampton, Mark Clark, and Malcolm X—as well as Denise McNair, Addie Mae Collins, Carole Robertson, and Cynthia

Wesley, the little girls murdered in the 16th Street Baptist Church bombing in Birmingham—are represented as cherubim, an allusion to the baroque period. The distinctive palette and composition, the black angelic figure with gold glitter wings, and the epic scale are all signature Marshall, firmly establishing the painting as a contemporary work of art. *Souvenir I* and all the paintings in the series are layered with multiple, intersecting histories: black and American history, the history of Western art, and contemporary history.

Marshall's work also incorporates elements of black folk art and symbols that have powerful significance in African traditions. Commenting on his 1990 painting *When Frustration Threatens Desire,* he said, "What also started to crystallize in that painting was a way to bring together not only Western traditions of pictorial representation, but folk traditions of painting that have an equally valid authority. I don't see much difference between using Giotto or Bill Trayler as a point of reference. To me they're the same. Their work has a certain power."

The allusion and symbolism in Marshall's work are more than artistic devices. Marshall uses *symbolism* in the fullest sense of the word—his symbols are not merely signs that stand in for something else. He uses symbol in a way that fills his art with a "certain *power.*"

The American theologian, poet, and activist Thomas Merton wrote, "The true symbol does not merely point to some hidden object. It contains in itself a structure, which in some way makes us aware of the inner meaning of life and of reality itself. A true symbol takes us to the center of the circle, not to another point on the circumference. A true symbol points to the very heart of all being, not to an incident in the flow of becoming."

In this sense, a true symbol is timeless.

Remembrance of the Future

How we keep time is itself symbolic. In physics, time is typically denoted by the symbol t; its relationship to distance, speed, energy, and power is reduced to formulas that strip it of its immaterial, mysterious quality. Our calendars and watches symbolize the flow of time and divide it into discrete elements like slices of bread. But what are these "seconds," "minutes," "hours," and "weeks" that maintain such strict rule over so many of our lives?

The way we keep track of time is an entirely human construct, though it has origins in both culture and astronomy. Our modern system consisting of the *year, month, week, hour,* and *second* is a sort of multicultural grab bag that reflects humanity's creative and varied methods of timekeeping.

Cultures throughout the ages have developed timekeeping systems based on cycles observed in the sky. The most intuitive way to divide and measure time is the day, defined by the rising and setting of the Sun or stars due to Earth's spin on its axis. To define longer periods of time, people have looked to the Moon, Sun, stars, and planets.

The 12-month calendar based on the lunar month is a common feature of many societies, including the ancient Romans and a number of present-day Asian cultures. Most lunar calendars are actually lunisolar. Since it takes 29½ days for the Moon to return to the same position in the sky, 12 months work out to only 354 days. So most lunar calendars insert extra days into the year in order to align with the 365-day solar calendar.

The ancient Egyptians created the first 12-month solar calendar based on the Sun's position with respect to the stars. Their civic calendar was grouped into three seasons, each containing four months, each month 30 days long, with every week 10 days. The extra five days at the end of the year were designated for special celebration. The simple elegance of their well-organized year stands in stark contrast to the Gregorian calendar, with its mishmash of 30- and 31-day months, not to mention February!

The ancient Maya had a complex system of three interconnected calendars. The *Haab* calendar, based on the 365-day solar year, consisted of 18 20-day months, plus an additional five-day month that had special spiritual significance. Each day in the month was named with a unique *Nahual*, a symbol representing a protective spiritual being that usually took the form of an animal. Within the *Haab* was the 260-day *Tzolkin* calendar, consisting of 13-day "weeks."

Societies around the world have found the concept of a "week"—a period of time longer than a day but shorter than a month—an attractive and useful way to organize themselves. A culture's definition of a week is loaded with meaningful symbolism that usually represents a "throwing together" (the root of the word "symbol" in Greek) of their natural environment, spiritual beliefs, and cosmology. In traditional Yoruba culture, for instance, the week has four days, each of which is dedicated to a principal god who helped create the universe.

Our modern seven-day week first arose in Babylonia, and in later periods it was also adopted by the ancient Jews, Chinese, Indians, Greeks, and Romans. The Babylonians may have chosen seven days because that is about how long it takes between each of the four most prominent lunar phases— new Moon, first quarter, full Moon, and third quarter. The ancient Romans put their own spin on things by naming the days after their important deities, which in their astrology were associated with the five planets visible with the naked eye, the Moon, and the Sun.

Over the millennia, humans have devised a multitude of ingenious devices for keeping time on intervals shorter than the day. From sticks or stones set into the ground—so that their moving shadows marked the day's progress—to sundials, candle clocks, and water clocks, humans have indulged their obsession with tracking time. So how did we arrive at 24 hours in a day? Why are hours and minutes divided into 60 parts? And where did our most basic unit of time, the second, come from?

Once again, we can credit the ancient Egyptians and Babylonians for part of the answer to these questions. The Egyptians were the first to subdivide the day and night into 24 parts. By 1500 BCE, they had invented a sophisticated sundial with 12 equally spaced increments that marked the time between sunrise and sunset. Keeping track of the 12 divisions at night involved a more complex system of observing certain sets of stars. The significance of the number 12 is related to the 12 lunar months of the year. The Babylonians were the first to use a base-60 counting system. The origins of the sexagesimal are unclear, but mathematically, we know that 60 is an incredibly

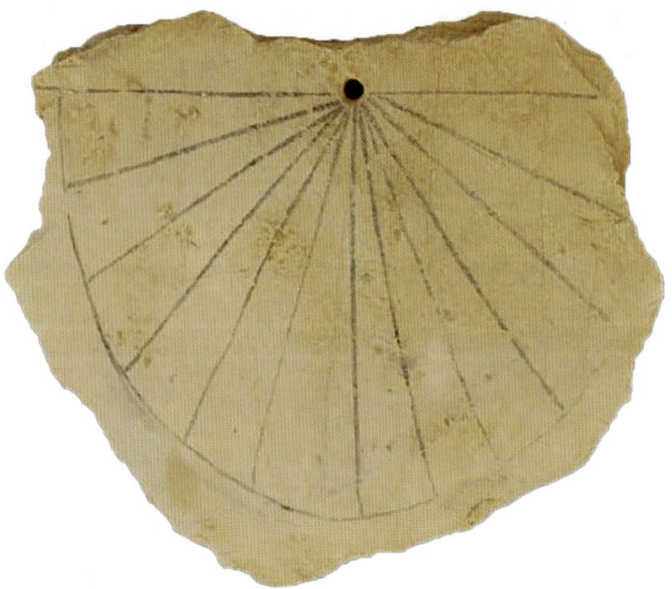

THE OLDEST DISCOVERED SUNDIAL | In 2013, archaeologists reported discovering this ancient timepiece, dated about 1500 BCE, in Egypt's Valley of Kings.

beautiful number. Sixty can be evenly divided by the first six counting numbers, as well as 10, 12, 15, 20, and 30, making it very convenient for dealing with fractions.

We can thank Al-Biruni, one of the most significant scientists and scholars of the Islamic Golden Age, for giving us the *second*. The prolific Persian polymath was concerned with time in many of its manifestations, from the phases of the Moon to the evolution of the cosmos. (He argued that the universe had a beginning, going against Aristotle's view that the universe was eternal and unchanging.) Al-Biruni was the first to split the hour into 60 minutes and the minute into 60 seconds. Thus, historically, the second is equivalent to about 1/86,400 of a day.

But in the early 1900s, it was discovered that the length of the day is getting longer by about 0.0014 seconds each century because tidal friction with the moon is gradually slowing down Earth's spin. Today atomic clocks are based on a much more precise definition of the second: "the duration of 9,192,631,770 periods of the radiation corresponding to the transition between the two hyperfine levels of the ground state of the caesium-133 atom."[*] In normal English, one second is the time that elapses during more than 9 billion vibrations of a photon produced by a particular atom.

Compared to other prominent scientists who did not surpass him in stature or influence, Al-Biruni's name is little known today in the West, but our modern world is synchronized according to his legacy.

Humanity has come a long way from naming the days after deities and spirits and charting the hours of the night with the stars. The way we think about time now often strips it of all its more spiritual and poetic connotations. Our phone and computer screens flicker lifelessly through the minutes, hours, and days—an inadequate way to express time's deeper symbolic meanings that resonate with the human heart—so that we, too, are in danger of becoming machines when we mindlessly obey them. Khalil Gibran reminds us of how fully human is our connection to time in *The Prophet*:

> And an astronomer said, "Master, what of Time?"
>
> And he answered:
>
> You would measure time the measureless and the immeasurable.
>
> You would adjust your conduct and even direct the course of your spirit according to hours and seasons.
>
> Of time you would make a stream upon whose bank you would sit and watch its flowing.
>
> Yet the timeless in you is aware of life's timelessness,
>
> And knows that yesterday is but today's memory and tomorrow is today's dream.
>
> And that that which sings and contemplates in you is still dwelling within the bounds of that first moment which scattered the stars into space.

---

[*] This definition of the second was given in 1967 by the Thirteenth General Conference of the International Committee for Weights and Measures.

Who among you does not feel that his power to love is boundless?

And yet who does not feel that very love, though boundless, encompassed within the center of his being, and moving not from love thought to love thought, nor from love deeds to other love deeds?

And is not time even as love is, undivided and paceless?

But if in you thought you must measure time into seasons, let each season encircle all the other seasons,

And let today embrace the past with remembrance and the future with longing.

Our modern, scientific, basic unit of time—the second—is based upon light. The 9.1-gigahertz frequency of radiation from the cesium atom corresponds to a wavelength of 3.26 centimeters, in the microwave portion of the electromagnetic spectrum. A photon having this precise wavelength is produced whenever the cesium atom's outermost electron flips over. Put more concisely—and lyrically—we measure time according to light's memory.

Light has memory. Whether it is captured in the colors of a canvas or raining on Earth from the planets and stars, light bears the history of another place in spacetime. By letting its remembrances unfurl before us, we can discover the inner workings of the universe and, through art, discover ourselves.

Art is perhaps the closest we will ever come to time machines. A painting or a poem is a crystallization of the artist's knowledge, emotions, thoughts, and intentions at the time of creation; it is meant to be experienced in the present; and, if it is effective, it carries within it the seed of a vision for the future. "The painter should not paint what he sees," said Paul Klee, "but what will be seen." The potency of a work of art can be judged by the degree to which it changes one. At the same time, the responsibility lies with the viewer, who decides how open and receptive she is to being transformed. If our heart is sensitive, the experience of a work of art can propel us, forever changed, into the future.

CHAPTER EIGHT

HARMONY OF THE WORLD

Thou didst erect the heavens themselves out of harmonies.
— Johannes Kepler —

In art, and in the higher ranges of science, there is a feeling of harmony which underlies all endeavor. There is no true greatness in art or science without that sense of harmony.
— Albert Einstein —

When they [paintings] are done right, harmony appears by itself. The more numerous and varied they are, the more the effect is obtained and agreeable to the eye.
— Paul Cézanne —

T hese murals were painted by Mr. Aaron Douglas of New York, an African American artist."

These are the words displayed on a sign posted at the entrance of the Hall of Negro Life at the Texas Centennial Exposition of 1936. The murals in question had been commissioned from a leading artist of the Harlem Renaissance for the Hall of Negro Life, a special exhibition that black leaders had fought hard to acquire federal funding for. After the exposition concluded, the Hall of Negro Life was destroyed—together with two of Douglas's four extraordinary murals—in spite of protests from the black community.

The Hall of Negro Life opened on Juneteenth. The sign had been put up after countless white visitors to the world's fair had repeatedly expressed disbelief that a black individual could have created the stunning site-specific murals. *Aspiration* was one of the four panels that Douglas painted portraying the history of Texas from a black perspective. The second surviving painting, *Into Bondage*, depicts captured Africans standing on the shore, about to be shipped across the Atlantic into slavery. *Aspiration* represents a later stage in history. Three confident figures dominate the canvas—two men and a woman. They embody the strength and perseverance of a people who have survived the horrors of slavery—still a living memory in 1936—invoked by the upraised shackled hands, but who have yet to achieve the promise of freedom symbolized by the dazzling city-on-a-hill in the distance.

Interlacing Harmonies

Aspiration captures and keeps the attention because it is rich with layer after layer of harmonies. We human beings love harmony. Our brains are wired for it. The five-by-five-foot painting is an object lesson in harmony, one of the principles of design in visual art. Harmony is the uniting of multiple different parts that together make up a whole that's greater than the sum of its parts. In music, harmony is the joining of two or more notes to create a new, usually pleasing sound. Analogously, in painting, harmony refers to how the various elements of art are combined and cooperate in a single work.

The seven elements of visual art are usually counted as color, value, line, shape, form, texture, and space. Harmony is achieved when one or more similar elements are repeated or arranged in patterns to create a sense of overall balance, rhythm, and unity. Effective, enduring works of art are layered with multiple harmonies yet have enough variety to sustain interest.

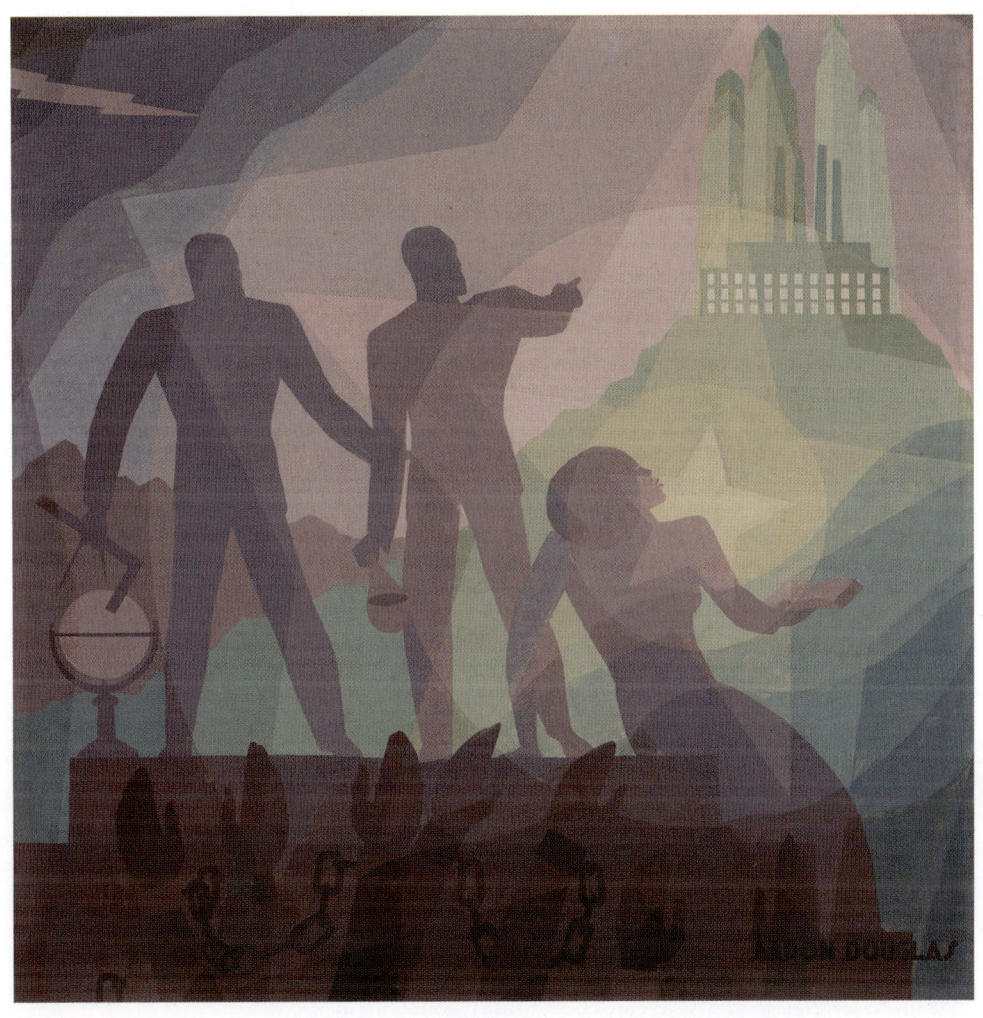

ASPIRATION | Aaron Douglas, 1936, oil on canvas, 60 × 60 in (152.4 × 152.4 cm).

© 2024 Heirs of Aaron Douglas / Licensed by VAGA at Artists Rights Society (ARS), New York. Fine Arts Museum of San Francisco.

Aaron Douglas's *Aspiration* attains harmony in several ways. The color harmony strikes immediately. The palette is dominated by cool colors—serene blues, pinks, and purples. The prevalence of purple has significance as the color of royalty in ancient Egypt, a recurring theme in this work. Douglas was working at a time when the public interest in ancient Egypt was at an all-time high. The discovery of King Tutankhamun's tomb in 1922 had created a media sensation, and artists and intellectuals of the Harlem Renaissance seized the moment to reclaim Egypt as the source of black history and culture. By linking black Americans to one of the world's oldest and most sophisticated civilizations, artists like Douglas were challenging the racist stereotypes that the mainstream media presented and replacing them with images of black people as uplifted, dignified, culturally refined, and intellectually distinguished.

Forms associated with ancient Egypt recur throughout *Aspiration*. The tiered platform raising up the woman and two men alludes to Egyptian architecture, while their silhouettes call to mind the iconic profiles often associated with ancient Egyptian art. The flatness of the figures, and the painting overall, is characteristic of Douglas's work, reinforced by his smooth, flat brushwork. Douglas was a student of other African art as well, and the woman's face, with the distinctive slits for eyes, is directly inspired by the Dan mask that Douglas would have seen as a frequent visitor to the Brooklyn Museum. Like masks created by people throughout the African continent, those made by the Dan of Liberia had spiritual significance, connecting those who wore them to unseen, supernatural forces. By linking modern black Americans to their African culture and ancient past, Douglas was signaling that their history predated their arrival in America as enslaved people.

The palette of *Aspiration* is limited to colors of the same *intensity,* or *saturation*—another way this painting achieves harmony. Yet there is variety, too, in the subtle gradation of hues, from yellows to blues and purples, and the contrast in value (light versus dark).

Then there is the repetition of shapes—the radiating stars and circles. The star is a symbol of the Lone Star State, but black visitors to the Hall of Negro Life would have also recognized it as the North Star, a symbol of freedom. The concentric circles, appearing in many of Douglas's paintings, are a motif he associated with music—Negro spirituals and jazz, in particular. The stars and circles radiate from the heart of the seated woman through the space of the painting, their light fracturing into a kaleidoscope of color as they intersect with the various forms and figures. The woman holds an open book, the symbol of knowledge, suggesting that wisdom, culture, and enlightenment have a feminine origin.

Yet another layer of harmony arises from the complex interaction of sharp, angular lines and soft, undulating curves, which might suggest the balance of masculine and feminine, science and art, or spiritual and worldly knowledge.

Douglas masterfully works each of the seven elements of art to create a richly interlaced superharmony that is greater than the sum of the individual harmonies composing it. The technical virtuosity of the painting serves an even higher purpose. In addition to references to African art and black American music, Douglas makes a number of other historical and cultural connections in *Aspiration*. His bold use of geometric shapes was influenced in part by innovations in modern art, including art deco and cubism. He also drew from his knowledge of earlier periods of Western art. Art historians suggest that the reclining pose of the female figure in *Aspiration* is inspired by Michelangelo's painting *The Libyan Sibyl*, which depicts an ancient prophetic figure. Douglas no doubt also had in mind a more recent prophet and guide. Sojourner Truth became widely known as the Libyan Sibyl after Harriet Beecher Stowe referred to her by that moniker in an article for the *Atlantic Monthly*. The standing male figures with scientific instruments in hand might be Benjamin Banneker and George Washington Carver, both men of faith who were committed to the advancement of their people. All three are paragons of black brilliance who made decisive contributions to the culture and used their talents in service of their people's liberation. They exemplify what is possible in terms of intellectual, moral, and spiritual excellence.

Well before the Texas Centennial, Douglas was already recognized as a leading voice of the Harlem Renaissance. He embodied the movement's collaborative spirit, as artists from all disciplines—painting, sculpture, photography, music, dance, literature—drew inspiration from one another and shared their knowledge. Douglas studied with the German-born artist Winold Reiss, who encouraged him to explore his African heritage and incorporate it into his art; he collaborated with Langston Hughes, W. E. B. Du Bois, and other writers; he supported young black artists through his work in the Harlem Artists Guild.

In the mid-1920s, James Weldon Johnson invited Douglas to collaborate with him on *God's Trombones: Seven Negro Sermons in Verse.* Johnson wrote the poems, patterned after the voice of the traditional black preacher. He wanted to celebrate this vital leader of the black community and display the high art of his preaching. "It was through him," Johnson wrote, "that the people of diverse languages and customs were brought here from diverse parts of Africa and thrown into slavery were given their first sense of unity and solidarity." Indeed, the black preacher was a manifestation of the cultural harmony of a diverse people unified by their aspiration for freedom and autonomy.

Douglas illustrated these sermons—poems with titles such as "Creation," "Let My People Go," and "The Judgment Day" that used stories from the Bible as allegories for the black experience. Douglas's signature style already contained many elements that would later characterize *Aspiration:* the bold geometrical shapes; strong, intentional lines; circle patterns; striking silhouettes viewed in profile. The physiognomies of the central characters—from Adam to Simon of Cyrene—show the influence of African masks and ancient Egyptian art. Most depictions of famous Bible stories familiar to American audiences of the time were portrayed with white people. In making these heroes black and using visual motifs that would resonate with black people, Douglas was inviting his community to take new ownership of a religion that has had a profound influence on black American culture.

Light is a powerful theme connecting the illustrations in *God's Trombones.* In each image, a ray of light or some other representation of light cuts through the space and illuminates the main subjects. This theme appears throughout Douglas's body of work. His figures often appear in direct contact with a lamp, the Sun, a star, or some otherworldly radiance emanating from beyond the image. In his symbolic use of light, with its implication of divine

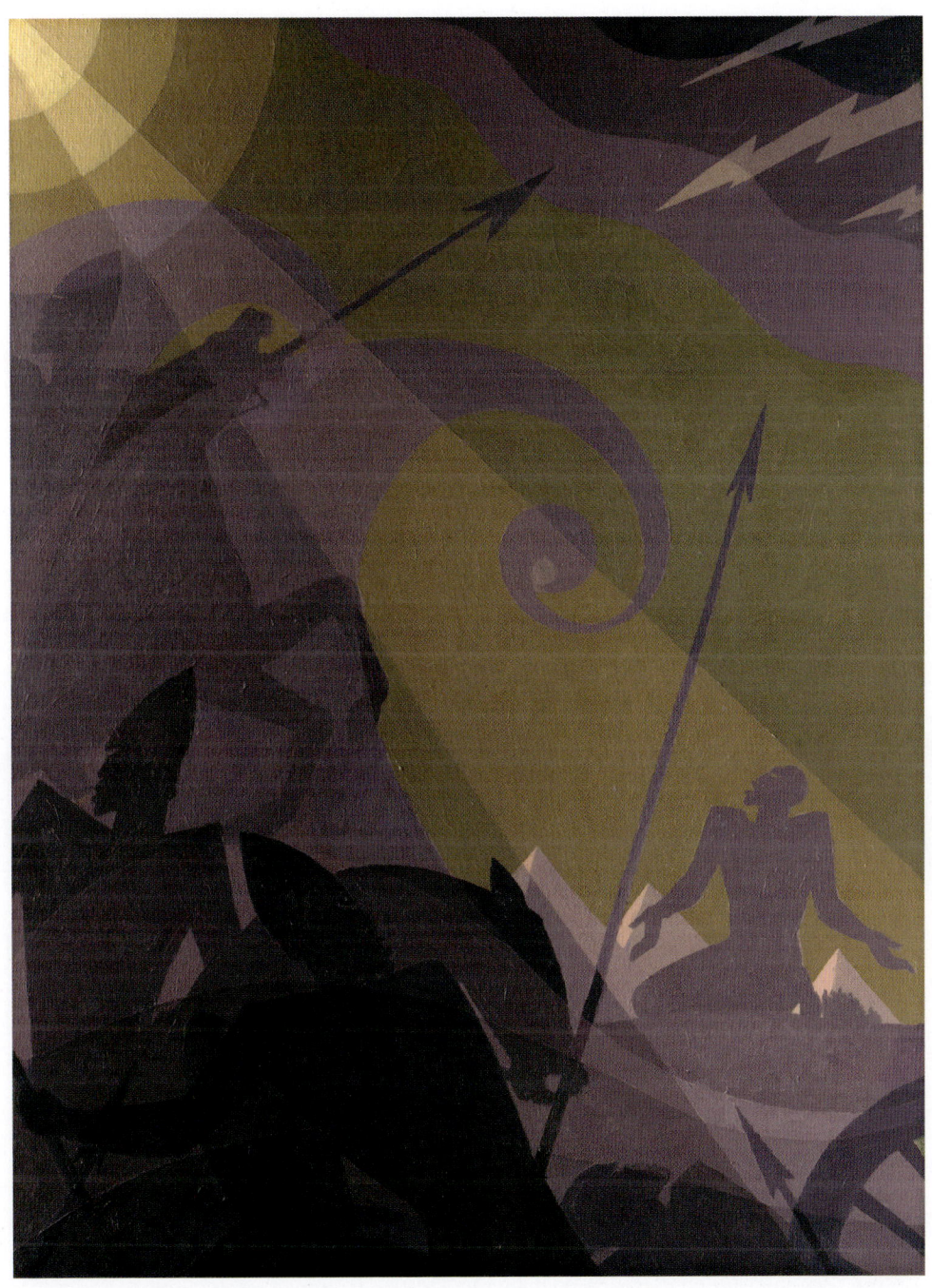

LET MY PEOPLE GO | Aaron Douglas, ca. 1935—39, oil on masonite, 48 × 36 in (121.9 × 91.4 cm).

illumination and of humanity's connection to the cosmos, light is a visual reminder that we are part of the same human family.

Aaron Douglas was working at a time when racist representations of blacks and other nonwhite people proliferated. His images reframe Western narratives that present people of European descent as the source of civilization and everything else of value in the world. Douglas's art helps restore balance in a culture severely lopsided by racism and white supremacy. The harmony realized in a painting like *Aspiration* is not only aesthetic but cultural and social. The artist used many of the technical, aesthetic harmonies—color, intensity, form—to serve a higher symbolic purpose. This was Douglas's genius. It is in this symbolic sense that the harmony of Douglas's art is most fully realized.

"The Earth Sings Mi, Fa, Mi"

Humans are metaphorical creatures. We think and speak in metaphors automatically. They help us to understand and make sense of life; they help make our highest aspirations feel more concrete and possible to achieve. Harmony is about joining things together in a cooperative way. In fact, "agreement of sounds," "union," or "joining"—in Greek, *harmonia*—is the original meaning of the word "harmony."

When we talk about "inner harmony" or "living in harmony" with one other, the metaphor here is *life is music*. In music, there are ways to bring notes together that sound good and feel right, as though they agree. There are ways to bring together notes that clash and create tension; they sound and feel terrible. The same with human relationships, the marrow of life. Our relationship to ourselves, to other human beings, to nature can be characterized by agreement and unity or discord and division.

The Chinese philosopher Confucius also linked harmony to music and considered it a virtue worth striving for. In his wisdom, he understood that diversity is essential to promoting harmony. Indeed, harmony involves the joining together and cooperation of disparate things. In *The Analects*, a book of his sayings compiled by his students, Confucius says, "The noble-person searches for harmony without uniformity, but the petty person searches for uniformity without harmony." You can't achieve harmony with sameness. To realize the wholeness, elevation, and rich beauty we all want, we need variety. The hallmark of disharmony is tension, disorder, incompleteness. We need harmony because it gives rise to balance, order, peace, and integrity.

Confucius taught that pursuit of inner and social harmony should extend to our relationship with nature. I understand this to mean that because we are part of nature, and nature—the universe—is literally in us, it benefits everyone when we seek to join in its harmony. Furthermore, if the universe is by definition everything that exists, this means that everything about us— not just our physical selves but all of our immaterial aspects, like consciousness, love, spirit, and soul—emerged from the universe. Leonardo da Vinci reflected, "Our soul is composed of harmony." I would add that we embody the harmony of the cosmos itself.

The universe makes its own sort of music, an idea that goes back to antiquity. In ancient Greece, philosopher-mathematician Pythagoras proposed that the motions of celestial bodies produce unique tones. His mathematical experiments with music led him to notice that plucking strings of different lengths produces different notes. He found that certain combinations, corresponding to simple mathematical ratios, sound good together. For instance, the ratio 2:1 represents an octave. In other words, if one string is twice as long as another, the sounds the two create will be an octave apart. If two strings have a 3:2 ratio—one string is two-thirds the length of the other—when plucked their pitches will form what musicians call a *perfect fifth*, the interval you hear when you strike the C and G keys on a piano. Pythagorean intervals laid the groundwork for many systems of tuning and harmonies in Western music.

Pythagoras sought to extend his theories about harmony to the heavens. He argued that the simple whole-number ratios that form pleasing musical intervals also apply to the planets and other celestial bodies, which the ancient Greeks believed revolved around Earth. The orbital motions of the Moon, Sun, and planets relate to their distances, which correspond to different lengths of string. "If earthly objects such as strings or pieces of metal make sounds when put in motion, so too must the Moon, the planets, the Sun, and even the highest stars. As these heavenly objects are forever in motion, orbiting the Earth, surely they must be forever producing sound." Pythagoras reasoned that each celestial body sings distinct notes based on its orbit. Together, he believed, this *harmony of the spheres* influences life on Earth. Why don't we perceive this celestial symphony? Since it plays continuously, Pythagoras contended, our ears have grown accustomed to it and don't distinguish it from silence.

Centuries later, the 17th-century German astronomer Johannes Kepler ran with the Pythagorean idea. But in contrast to the ancient Greeks, who believed that Earth stood at the center of the universe, Kepler used detailed observations to prove that all the planets orbit the Sun. The Greeks also believed the planets moved in perfectly circular orbits, which would imply that they always move at the same speed. But Kepler showed that the planets actually move along elliptical paths, which means that their speeds change as they orbit the Sun.

Kepler did not, however, abandon the idea that the planets sing and that their song affects conditions on Earth. A deeply spiritual man, he believed that God speaks to us through the music of the spheres. His religious faith was a huge motivating factor in his devoted scientific study of planetary motions. He published his groundbreaking discoveries about how planets orbit the Sun in *Astronomia Nova* (*New Astronomy*, 1609) and *Harmonices Mundi* (*The Harmony of the World*, 1619), works that in their own right form unique harmonies of astronomy, mathematics, theology, and philosophy.

Kepler found that the distances between the planets and Sun do *not* follow simple harmonic ratios. Undiscouraged, he reasoned that the relationship between a planet's maximum and minimum speed is the more suitable comparison for musical intervals. He converted the patterns he observed in the motions of the planets into tones. For instance, the maximum and minimum speeds of Mars represent a 3:2 ratio, a perfect fifth, while Saturn sings out a major third (5:4), and Jupiter, a minor third (6:5).

Observing that the maximum and minimum speeds of Earth form a ratio that approximates a semitone (16:15), corresponding to *mi* and *fa* in solfège,

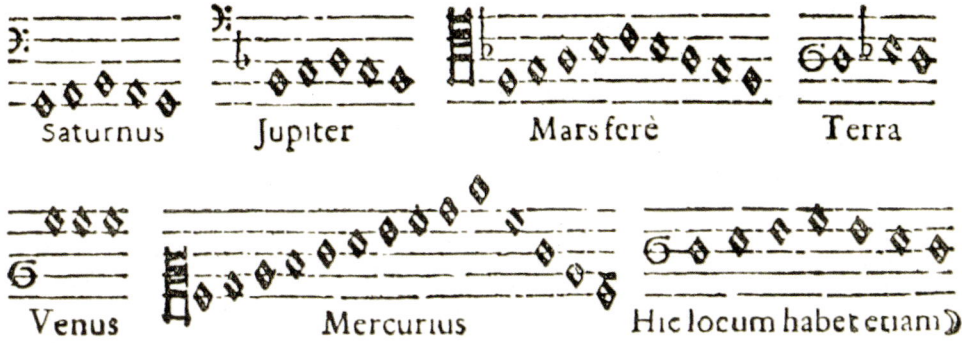

Saturnus   Jupiter   Marsferè   Terra

Venus   Mercurius   Hic locum habet etiam ⟩

PLANETARY MUSIC | From Kepler's *The Harmony of the World*, 1619.

Kepler saw in this confirmation of a heavenly choir singing out humanity's condition and fate. "The Earth sings Mi, Fa, Mi—so we can gather ever from this that *mi*sery and *fa*mine reign in our habitat." An original, open-minded thinker in many respects, Kepler nevertheless singled out Latin, in which he wrote *Harmonices Mundi*, as God's chosen language ("misery" = *miseriae*; "famine" = *fames*), and the chromatic scale used in Western music as God's preferred basis for composing. It may be tempting to laugh, but scientists today are no less imbued with the assumptions and values of their culture.

Whether or not we accept the idea that the planets sing in Latin, or even Greek, is there any truth to Kepler's and Pythagoras's concept of *harmony of the spheres*? As already noted, Kepler's finding that the distances between individual planets and the Sun *do not* produce simple harmonic ratios pretty much demolished that idea. But Kepler was on to something in his belief that any hypothetical celestial music would be inaudible. Unlike Pythagoras, he held that the planets' voices were "to be perceived by the intellect not the ear." Sound needs a medium—like the molecules in air or water—through which to travel. But in rarefied space, particles are so few and far between that the sound from singing planets would have virtually no medium through which to reach us. Even the sound produced by the Sun, as noisy as it is, essentially stays trapped at its surface.

Nevertheless, in the mid-1990s, professors at Yale University found a way to help us imagine what the planets might sound like, based on the ideas of Kepler. Professors Willie Ruff, a musician, and John Rogers, a musician and scientist, took Kepler's calculations of planetary speeds and, with the assistance of computer scientists, converted them to sounds that humans can hear. To do this, they sped up the tones for the six planets known to Kepler, by scaling one year down to five seconds. This process is analogous to what astronomers do all the time when they take x-ray, radio, or other astronomical observations invisible to the human eye and translate them into the visible range of colors the human eye can see.

Ruff and Rogers produced a recording of synthesized music that includes voicings for the six planets Kepler knew of, as well as Uranus, Neptune, and Pluto, which weren't discovered until 1781, 1846, and 1930.* The recording, titled *The Harmony of the World: A Realization for the Ear of Johannes Kepler's*, can be found online. It's an odd nine-part harmony of whistles, warbles, sirens, thumps, and pulses. Each planet's voice is based on the extent and varying speed of its orbit. Mercury, with an orbit consisting of 88 Earth days, whistles a rather annoying high-pitched tune. Jupiter, more than 13 times farther from the Sun than Mercury is, produces deep bass tones. The three outermost worlds, whose orbits are simulated as beats rather than pitches, provide the rhythm section.

The ancient Greeks believed the planets moved in circles at constant speeds, which would correspond to one note per planet and a single chord. Kepler came along and showed that planets actually move along elliptical paths, which means their speeds change as they orbit the Sun. Conceptually, this means that their (hypothetical) voices continuously modulate, and the overall harmony is much more dynamic.

We should keep in mind that the concept of the *harmony of the worlds* or *music of the spheres* is a metaphor, a beautiful mental image that speaks to the balance and elegant orderliness of the Solar System. Like all metaphors, it has limitations. The distances of the planets to the Sun *don't* form ratios exactly comparable to harmonic intervals in Western music, the way

———————————

* At the time this recording was made in the 1990s, Pluto was still considered a planet. In 2006, Pluto was recategorized as a "dwarf planet."

Kepler had hoped. Certainly, the planets don't sing in audible voices the way Pythagoras imagined. Yet as a metaphor, the *harmony of the worlds* does expose an important kernel of truth: the Solar System has its own special kind of harmony without which we might not be here. At the least, life would be very different.

In harmony, the various parts make up a synergistic whole. If one of the parts disagrees with the others or is missing altogether, things might be off or incomplete. What if the other planets didn't exist? How does the harmony to which they each contribute affect Earth? I'm not referring here to astrology—which astronomers consider a pseudoscience—but to the planets' physical influence on us and on our understanding of nature.

Isn't it nice that our planet isn't bombarded with asteroids more often? We can thank Jupiter for that. Although the gas giant sails around at a remote 484 million miles from the Sun—more than five times farther than Earth—its existence is critical to our own. Jupiter is the most massive planet in the Solar System, and its huge gravity often protects Earth and the other inner planets by deflecting asteroids and comets that might otherwise slam into them. The early Solar System was a chaotic place, and collisions between asteroids and the newly forming planets were much more common then. Such early impacts may have helped deliver water to Earth. Others were catastrophic, such as the Chicxulub impact that wiped out the dinosaurs and 70 percent of other species on Earth 66 million years ago.

As the second largest planet in the Solar System, Saturn has played a role similar to Jupiter's in sometimes shielding the inner planets from collisions. And the ringed planet affects life on Earth in another major way. Its size and position in the Solar System help stabilize Earth's orbit around the Sun. All the planets interact gravitationally, influencing one another's orbits. Saturn has a particularly strong effect on Earth. If Saturn were smaller, or if it were just 10 percent closer to the Sun, Earth's orbit would become more elongated. Right now, Earth's orbit is nearly circular, and throughout the year it remains comfortably within the habitable zone—the ring around a star where the temperature is just right for liquid water to exist on the surface of a planet. If Saturn's size or distance to the Sun were reduced, Earth's orbital path could

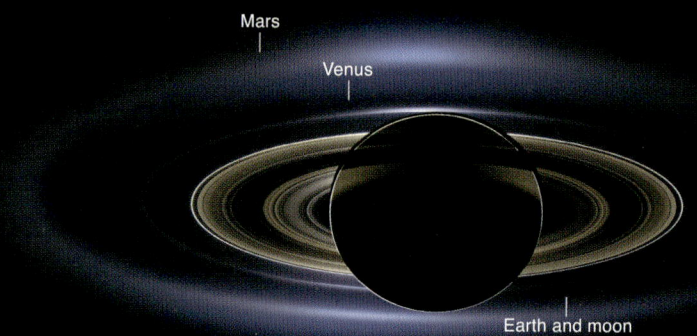

SATURN | In this remarkable 2013 image from NASA's Cassini spacecraft, Saturn and its inner rings are phantomlike as the Sun illuminates them from behind. Mars, Venus, and our own planet appear in the background as nearly imperceptible specks of light.

pass out of the habitable zone for months each year, affecting the supply of liquid water and the life that depends on it.

Could we do without our twin planet? Since antiquity, people have admired Venus as both the "morning star" and "evening star." Venus is always rather close to the Sun in the sky. With its thick, reflective clouds composed mostly of droplets of sulfuric acid, it is bright enough to be easily spotted just after sunset or a few hours before sunrise. For about 263 days, when Venus leads the Sun in the sky, it is visible in the morning. It then disappears behind the Sun for about 50 days before reappearing as the evening star, at which point it will trail the Sun for another 263-day period.

The cycles of Venus formed the basis for one of the Mayan calendars, and the planet has inspired art, poetry, and folklore across many cultures. Venus also plays an important role in our scientific understanding of the universe. In 1619, Kepler calculated the relative distances of the planets to the Sun. Then in 1716, the English astronomer Edmond Halley—after whom Halley's Comet was named—suggested a method to calculate the distance to the Sun itself, using the transit of Venus.

Because the Sun appears as a circle on the sky, when Venus crosses in front of it, the length and duration of its path can be measured. Transits of Venus, occurring just every 243 years, are very special for astronomers. The last pair of transits occurred in June 2004 and June 2012. The next pair won't take place until December 2117 and 2125.

The angle of Venus's path across the Sun and the time it takes to transit appear different for observers located in different places on Earth. By comparing the paths—and using knowledge of Kepler's laws of planetary motion, plus some simple trigonometry—calculations can be made to infer Earth's distance to the Sun: about 93 million miles. Nowadays, we calculate the exact distances to planets using radar. But this method doesn't work for the Sun because it generates lots of radio waves, which drown out any signal from radar reflection.

All of the planets contribute to humanity's experience of the cosmos, and we can only imagine how different life would be—culturally, artistically, scientifically, and biologically—if any one of them weren't here. Though the

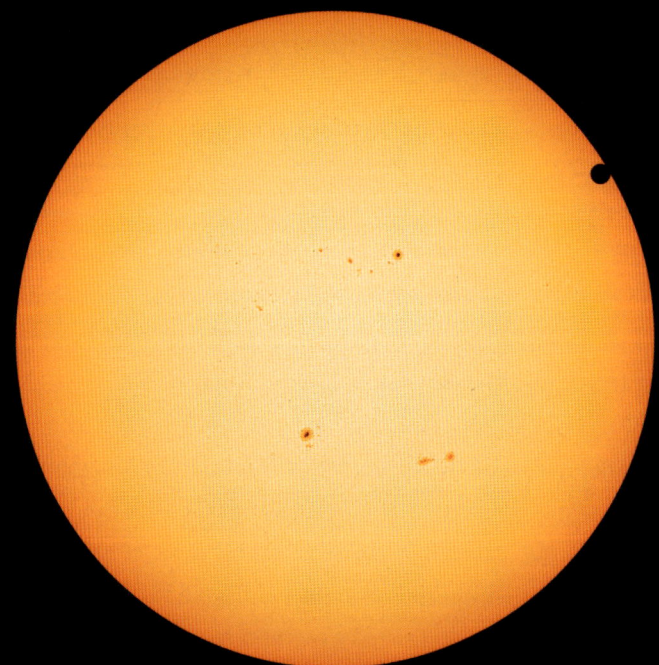

VENUS CROSSING THE SUN | Transits of Venus are rarer than total eclipses of the Sun by the Moon. This image, taken by NASA's Solar Dynamics Observatory during the 2012 transit, captures Venus just as it passed from the face of the Sun.

planets don't exactly sing, their harmony makes life possible. As it turns out, we also inhabit the *literal* harmony of the entire universe.

The "Big Silence"

Sound is a physical reality of the natural world, a vibration that travels as a wave through a medium. Any disturbance that causes vibration is capable of producing sound. In the physics definition of the term, it doesn't matter whether the sound waves are audible for them to be real. We are immersed in harmonies we cannot hear. And in fact, the universe was molded by sound.

The vast, empty regions between planets and stars prevent sound from traveling through space. But this wasn't the case in the infant universe. In the early years after the Big Bang, long before stars and galaxies came into existence, the universe was a roiling plasma, a sizzling gumbo of electrons, protons, and light. Space was much denser than it is now—denser than air on Earth, in fact—so that sound waves could propagate through it. For hundreds of thousands of years, the entire cosmos hissed. Had anyone been around to listen, they might have been disappointed, for the hiss was some 50 octaves lower than what the human ear can perceive. Contrary to the name *Big Bang*, our universe was birthed in silence.

These hissing sound waves were induced by extremely tiny fluctuations appearing immediately after the Big Bang. As these primordial sound waves traveled through space, they left their imprint on light and matter, similar to how stones thrown into a pond send ripples across its surface. As the ripples on a pond are areas where more water has piled up, in the infant universe, the ripples were regions where the light got brighter and more matter piled up. As the universe expanded and cooled, the sound waves lengthened, and the hissing got lower and lower, until eventually it faded away altogether. Yet millions of years later, as the universe continued to grow, these ripples would become the places along which galaxies formed. Today we inhabit the echo of the cosmos's silent song.

If we could have observed the newborn universe from a bird's-eye view during the first moments after the Big Bang, the overwhelming sameness of it all might have left us feeling discouraged. We might have been hard-pressed to see how what appeared to be a uniform brew of matter and light might develop into the galaxies, stars, planets, continents, cities, and people that make up our reality.

What *appeared* to be.

According to Big Bang theory, in the first fractions of a second of the universe's life, matter and antimatter existed in *nearly* equal parts. Particles of matter and antimatter were spontaneously popping into and out of existence. Whenever a particle met its antiparticle, they annihilated each other, releasing pure energy—light. This could have gone on indefinitely, yet somehow, the antimatter particles were outnumbered by a hair's breadth. For every billion antiparticles, there were one billion *and one* particles. Within a second of the universe's birth, all the antimatter was wiped out, leaving behind a trace of matter. This tiny bit of extra matter survived and seeded the first elementary particles—neutrinos, electrons, quarks, protons, and neutrons—that would go on to build up our wonderfully complex world.

Today the structure of our universe is a sparkling cosmic web of galaxies. Had it been perfectly homogeneous in the beginning, the process of matter and antimatter annihilating each other would have continued much longer, the universe would have been filled mostly with a bath of light, and it's unlikely that life as we know it would have evolved.

Diversus

Why are we here? What explains that one-part-in-a-billion extra bit of matter present right after the *Bang*? Why was it just that amount? No more, no less. The reason for the matter-antimatter asymmetry remains one of the deepest mysteries in astrophysics. What we do know is this: We wouldn't be here without it. We exist because, somehow, diversity flickered into being in the infant universe, thus creating conditions for the emergence of harmony.

If understanding, peace, order, and integrity are sincere aspirations of humanity, nature and art are available to us as models for how to achieve them. Sameness gives birth to nothing. Evolution mandates diversity. And diversity makes harmony possible.

I don't want to be misunderstood. I don't mean this flabby notion of diversity that gets thrown around nowadays in every college, university, workplace, and business. A superficial "diversity" that struggles just to tolerate difference while never really threatening the structures of power that are the source of inequality and exclusion. A diversity that is justified not by its moral correctness but by its potential to increase productivity and profitability. In terms of metrics like race and gender, many institutions and organizations are more diverse than they've ever been, yet they lack overall integrity and by design remain inherently exclusionary.

I'm more interested in a diversity related to one of its roots in Latin—*diversus*, meaning "turned different or opposite ways." In this sense, diversity doesn't simply mean adding a bunch of different ingredients—race, gender, class, sexual orientation, religion, political ideology—just for the sake of a more colorful, equitable mixture that is more "representative," while the same people remain in power. It means turning in a completely different direction from the path we've been on; it means making a fundamental, one-part-per-billion change in course.

If harmony requires diversity, this means that harmony requires respect. A deep respect for life and for the unique, inherently valuable gifts that each individual bears within. In music, art, nature, and life, harmony presupposes unity and the active partnership of all elements involved. Harmony is diversity in cooperation.

CHAPTER NINE

EARTH QUILTED TO SKY

The only wisdom we can hope to acquire
Is the wisdom of humility: humility is endless.
— *T. S. Eliot* —

Science progresses best when observations force us to
alter our preconceptions.
— *Vera Rubin* —

Elizabeth Talford Scott loved the stars. Recalling her youth growing up on a former plantation in Chester, South Carolina, she said, "These stars back home were very precious to me. They gave us so much light. They lit our way home at night. They lighted up the porch. They even seemed to give off heat and warm us."

Scott remembered evenings when the family would sit on the porch, after the adults had finished a long day of work in the fields. She stitched these childhood memories of the starry sky into fabric, in the form of a quilt. The granddaughter of a former slave, Scott also pays homage to map quilts that enslaved blacks created to plan escape routes. The constellation of stars is surrounded by intricate embroidery, which Scott's daughter said represents a topological map of the plantation. "My mother was told . . . that slaves would work out a quilt piece by piece, field by field, until they had an actual map, an escape route. And they used that map to find out how to get off the plantation." This means that the *Plantation* quilt combines two perspectives into one: sitting among the stars and taking in an aerial view of the plantation below.

What moves me about this story is how Scott and her ancestors felt so connected to the stars. I imagine my own ancestors felt this way, too. They understood that there was always a higher perspective from which to consider their lives. That by the power of their intellect and imagination, they could rise toward this higher perspective, trusting that it would lead them to freedom. This required humility, a word whose meaning derives from "earth" and "on the ground." As Saint Augustine said, "Begin by descending. You plan a tower that will pierce the clouds? Lay first the foundation of humility." Certainly, no one was forced lower than the enslaved blacks of antebellum America. But their humility was not just a position they were forced into; it was an expression of free will that represented the determination to survive and achieve liberation.

When I think of Elizabeth Scott's story and others about how our people quilted the stars, the Western chronicle of humanity's relationship to the cosmos presents a curious contrast. Rather than "humility," words like "egoism" and "anthropocentrism" come to mind. With some notable exceptions, the history of Western astronomy is the history of humanity constantly being displaced from the center of the universe. It makes me think about how an attitude of humility very much depends on one's perspective.

The Art of Seeing

For nearly 1,500 years, Aristotle's belief that Earth sits still and the universe revolves around it dominated Western science. Ptolemy, an Egyptian mathematician, astronomer, and theologian, was born half a century after Aristotle. He adopted the orthodox ideas of his time and built a theory to explain

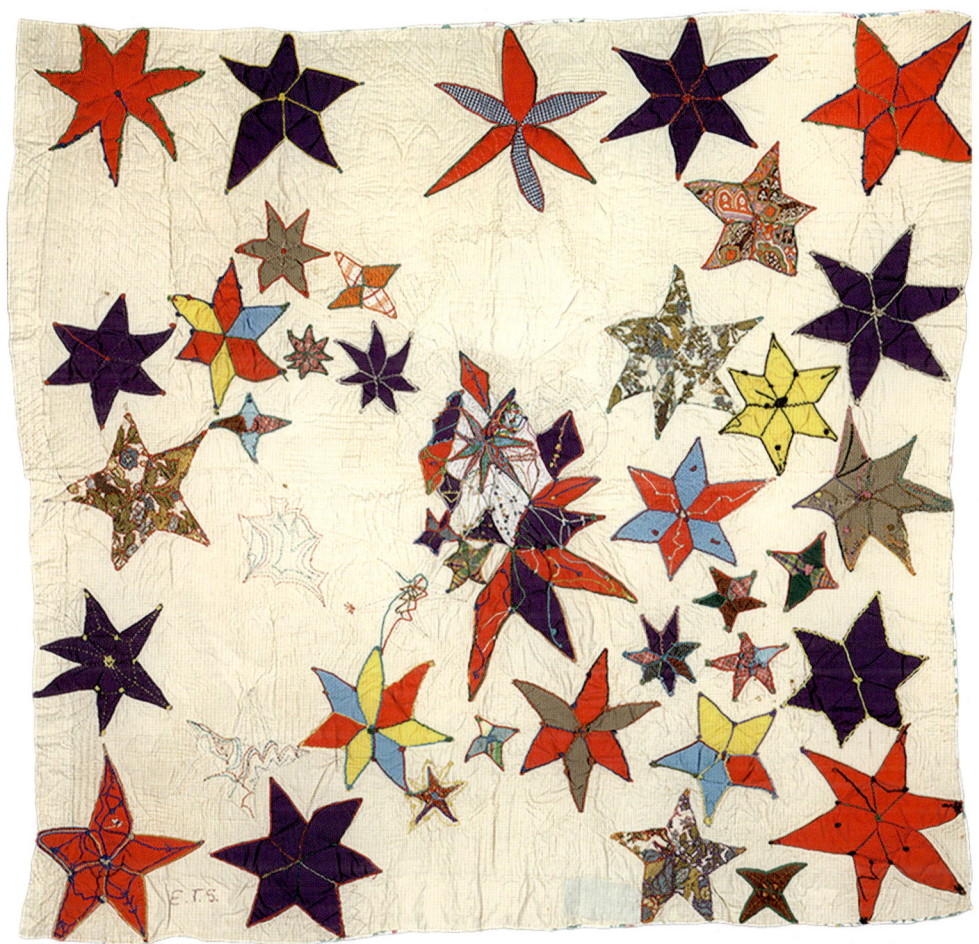

PLANTATION | Elizabeth Talford Scott, 1980, cotton, wool, and synthetic blend fabrics; cotton and silk threads; 68½ × 74½ in (174 × 189.2 cm).

© *Estate of Elizabeth Talford Scott at Goya Contemporary/TALP.*

the motions of celestial bodies. In Ptolemy's geocentric universe, Earth was imagined to be in the center of a series of transparent spheres to which the Moon, planets, and Sun were attached, with the stars fixed to the outermost sphere. Each celestial body moved along its sphere in a perfect circle, which, according to the ancient Greeks, was the most perfect and divine of shapes.

A few hundred years before Ptolemy, the astronomer Aristarchus proposed a model of the universe with the Sun at its center. Ptolemy rejected this idea, arguing that if Earth were moving at the rapid speeds required in this model, "The animals and other weights would be left hanging in the air, and the earth would very quickly fall out of the heavens. Merely to conceive such things makes them appear ridiculous." It's easy to smile at this now. But since he lacked our modern understanding of gravity, his objection that anything not pinned to the ground would be flung off into space did not seem so far-fetched to many. Ptolemy laid out his ideas and the theological underpinning for them in the *Almagest*, his most influential work. The geocentric model prevailed among European and Arabic astronomers well into the 17th century.

But an Earth-centered model of the universe still left a lot to be desired, as it was complicated by so many moving parts. The planets caused the biggest problem for geocentrism. In order to make sense of their motions, special caveats had to be included in Ptolemy's model.

The planets usually drift from west to east with respect to the stars. (Viewed from the North Pole, most everything in the Solar System spins and orbits in a counterclockwise direction.) But sometimes planets temporarily drift from east to west, an apparent backward movement known as *retrograde motion*. Due to its proximity to Earth, Mars presents the most obvious case. Every two years, Mars retrogrades for about two to two and a half months. Since Mars is one of the planets visible to the naked eye, Ptolemy was aware of this phenomenon and had to find an explanation for it that fit his theory. According to his Earth-centered model, Mars moves along a small subcircle called the *epicycle*, which spins around while attached to the larger rotating sphere, resulting in the apparent backward motion of the planet against the background of distant stars. This was already complicated enough, but a

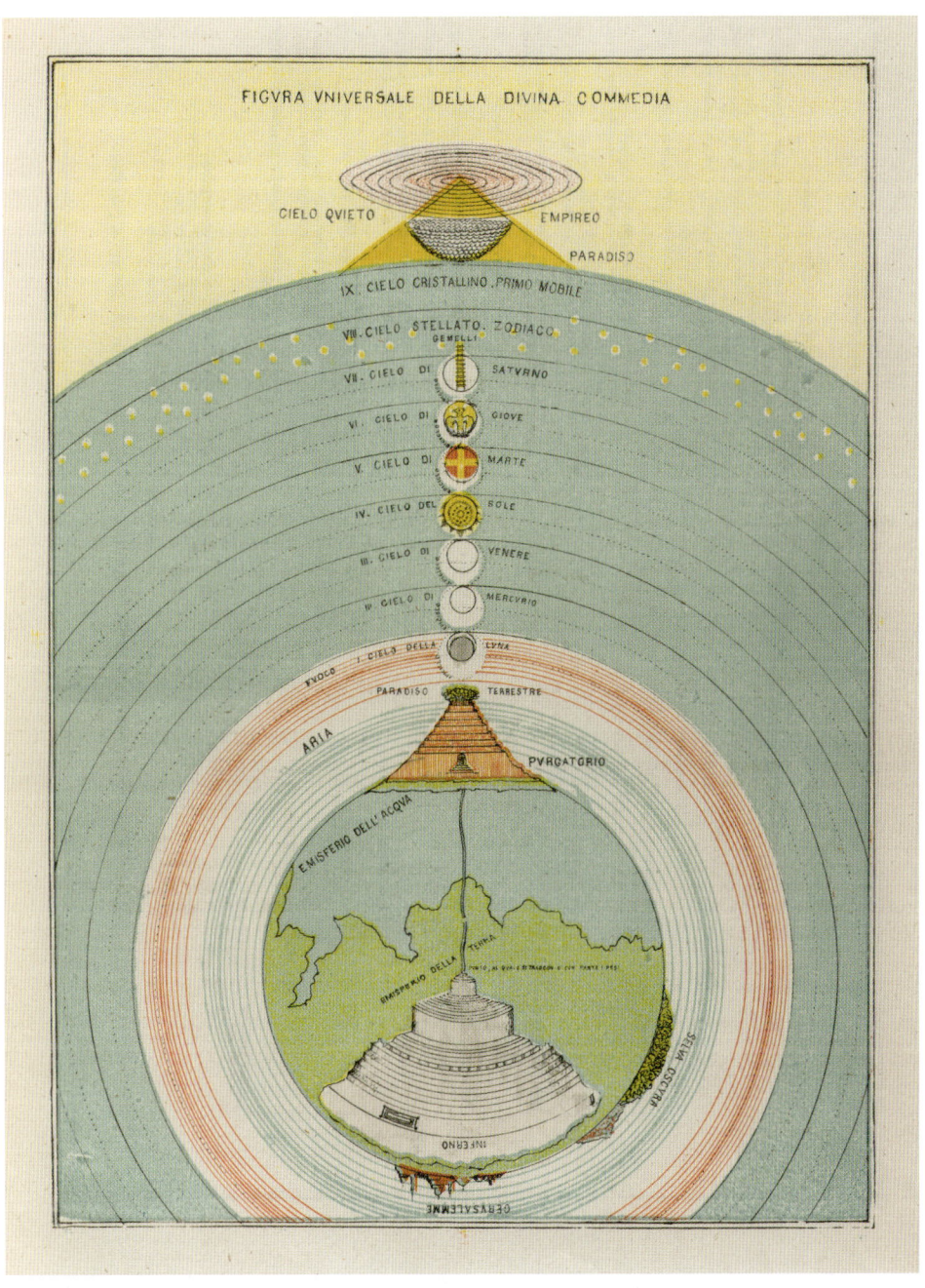

OVERVIEW OF THE DIVINE COMEDY | Michelangelo Caetani, 1872, lithograph, 9⅗ × 6⁷⁄₁₀ in, (24.5 × 17.5 cm). Ptolemy influenced not only scientists and philosophers, but artists, including Dante Alighieri, whose conception of the cosmos in *The Divine Comedy* (1321) was inspired by geocentricism. This lithograph is one of six maps designed by scholar and politician Michelangelo Caetani to aid students of Dante.

further adjustment was needed. Earth actually had to be slightly offset from the center of the nested spheres in order to make the geocentric model work.

By the early 16th century, a Polish Catholic canon had become convinced that Earth was not at the center of the universe and the theory that had held for more than a thousand years was wrong. Born in 1473, Nicolaus Copernicus studied to go into the clergy, but he also spent time in school pursuing math and astronomy. His research and independent observations led him to believe that the Sun is at the center of the universe, while Earth is just one among several planets orbiting it. The apparent circling motion of the Sun and distant stars around Earth could be explained by assuming that Earth itself is spinning.

Copernicus's heliocentric model provided a much simpler, more elegant way of explaining observations of movements in the sky, including the motions of the planets. In a Sun-centered universe, the correct ordering of the six known planets—Mercury, Venus, Earth, Mars, Jupiter, Saturn—fell easily into place. Copernicus correctly suggested, moreover, that the closer a planet is to the Sun, the faster its orbital speed. The simple reason for Mars's retrograde motion emerged naturally. As Earth overtakes the Red Planet in its orbit, the latter appears to go backward with respect to background stars. It's like when you're in a car that passes a slower moving vehicle: that vehicle appears to move backward.

Because Copernicus still held to the idea that spheres and circles were the preferred shapes of the cosmos, he had to rely on epicycles to explain the non-uniform speeds of planets. Though his model was flawed, it was an enormous improvement over geocentrism. Yet when he arrived at the same conclusion that Aristarchus did—now armed with observational evidence that the ancient Greek didn't have—he was going against the weight of a thousand and a half years of Aristotelian dogma. Perhaps fearing reproach from the scientific and religious communities, Copernicus didn't permit the publication of his masterpiece, *De revolutionibus orbium coelestium* (*On the Revolution of the Heavenly Spheres*), until just before his death, in 1543. Despite his reservations, he set in motion a revolution of another kind.

It would take many years for the educated of Western society to fully accept heliocentrism. It wasn't simply a matter of the logic of the theory needing to take root; people also needed time to allow the religious and philosophical implications to sink in. Remarking on the profound psychological impact in the Western world of what came to be known as the Copernican Revolution, Goethe wrote, "Of all discoveries and opinions, none may have exerted a greater effect on the human spirit than the doctrine of Copernicus. The world had scarcely become known as round and complete in itself when it was asked to waive the tremendous privilege of being the center of the Universe."

A big reason why heliocentrism wasn't immediately embraced is because astronomical observations had to catch up with Copernicus's theory. As often happens in science, when a new theory is presented, the technology needed to prove or disprove its implications lags behind. It may take years to gather enough evidence to firmly establish the theory. Much later, this happened with Einstein's theory of general relativity.

In November 1572, nearly three decades after Copernicus died, a new star suddenly appeared in the constellation of Cassiopeia. It was spotted by an eccentric Danish astronomer who, several years back, had lost a chunk of his nose in a sword fight and now wore a brass prosthesis to cover the unsightly wound. This unfortunate event didn't seem to diminish his self-confidence. By most accounts he remained flamboyant, stubborn, argumentative, and certain of his infallibility. Night after night, Tycho Brahe monitored the new star carefully. Initially brighter than Venus, it remained visible to the naked eye for well over a year before fading away. Some of Brahe's contemporaries still believed in Ptolemy's picture of the universe, and they supposed the object was very close to Earth—within the orbit of the Moon. This was because, according to the classic Greek view, the distant heavenly firmament should be unchanging. But Brahe wouldn't be swayed from this conclusion: the star had to be much farther away, because according to his meticulous observations, its position, unlike those of the Moon and planets, didn't change with respect to the stars. Today we know that the new star Brahe and many others around the world observed in 1572 was a supernova located at least 6,000 light-years away.

Brahe went on to design a major astronomical observatory that was built on an island gifted to him by the king of Denmark. With characteristic flair, he named it Uraniborg after the Greek muse of astronomy and supplied it with an array of instruments of his own invention, which he and his assistants used to chart the positions of the stars and planets with unparalleled accuracy. Operating during the last days before the invention of the telescope, Uraniborg was one of the first modern observatories. Although Brahe wasn't a follower of Copernicus, the exquisite data he collected over the course of 20 years would end up bolstering heliocentric theory.

Sadly, he had to leave his beloved island after falling out of favor with the king. He continued to practice astronomy in Germany and hired a young assistant by the name of Johannes Kepler. If Brahe was one of the first Westerners to take a modern approach to observational astronomy, Kepler was much more the theorist. He closely examined the data acquired by Brahe, who died just a year after they began working together, mining it for patterns and using math to explore, test, and develop his ideas. His years of work led to an undeniable conclusion: all evidence pointed to the Sun, not Earth, being at the center of the universe as it was then understood.

Kepler famously codified this model in his three laws of planetary motion. With Brahe's years of observations of the planets' motions at his disposal, he discovered general rules about their orbits around the Sun. Kepler's first law says that planets move in elliptical orbits, not circular ones. (An ellipse is sort of an elongated circle.) This means that as a planet sweeps around the Sun, its distance to it is constantly changing. This insight destroyed the idea that the heavens are full of perfect circles and did away with the need for epicycles to describe the motions of the planets.

Kepler's second law states that planets move faster when they are closer to the Sun. For the Earth, whose orbit is *nearly* circular, the change in distance to the Sun is marginal and its speed is fairly constant. Mercury has the most eccentric path of all the planets. As it orbits the Sun every 88 days, it gets as close as 29 million miles—speeding up as it draws near—and as far away as 43 million miles.

Kepler's third discovery says that a simple mathematical relationship exists between the *time* it takes for a planet to complete one trip around the

Sun and its *average distance* from the Sun. This elegantly simple rule fed into his belief that the planets sing together in heavenly harmony, a reality so profound in his mind that he named the book in which he published this third law *The Harmony of the World* (1619).

In Kepler's time the boundaries between various disciplines in the sciences and humanities were more fluid than they are now; they hadn't yet hardened into their current forms. With interests that included math, astronomy, and optics, Kepler was also motivated by philosophical and religious concerns; like his benefactor Tycho Brahe, he practiced both astronomy and astrology, still considered the same field of study at the time. For Kepler, a Sun-centered universe was not a threat to Christianity but an affirmation of his faith; he saw nature as mirroring the spiritual truth that God is at the center of life. The man who did so much to crystallize the parameters of modern Western science was far from rigid in his approach to what science is and is not.

While Goethe, an atheist, saw in Copernicus's heliocentrism a blow to humanity's egotism, Kepler believed that the elegant geometry of a Sun-centered universe was evidence of a "Divine Mind" for whom human beings are the purpose of creation. In *The Harmony of the World*, he writes, "Since geometry is co-eternal with the Divine Mind before the birth of things, God Himself served as His own model in creating the world (for what is there in God which is not God?), and He with His own image reached down to humanity." Belief in God and even the idea that people were created in "His own image" does not preclude humility with regard to humanity's place in the universe. Humility is a virtue in most of the world's religions and is not the exclusive claim of either secular or spiritual traditions.

The invention of the telescope in 1608 expanded our view of our place in the cosmos and transformed astronomy. It was like a curtain had suddenly been yanked off an enormous window that opened onto a universe much bigger, more dynamic, and more fascinating than anyone anticipated. It must have been an exciting time for the small group of astronomers who were essentially building themselves new eyes and, like newborns, were peering through them into a new world for the first time.

Originally the telescope was used for military activities and other Earth-bound purposes. The Italian scientist Galileo Galilei, a contemporary of Kepler, was one of the first to point his homemade telescopes heavenward. In January 1610, he observed four moons orbiting Jupiter through his eyepiece, proof that not all celestial bodies revolve around Earth. What he saw were the largest moons belonging to the gas giant—Io, Europa, Ganymede, and Callisto.* Since then, much more powerful telescopes have revealed as many as 95 smaller moons orbiting Jupiter, which comes in second only to Saturn, which has at least 145!

Galileo also saw firsthand that the heavens are full of irregular shapes—celestial objects and arrangements that don't conform to ancient Greek notions of perfection. He observed mountains and valleys on the face of the Moon, sunspots (he didn't know that looking at the Sun is dangerous), and strange "ears" on Saturn. A few decades later, Christiaan Huygens came along with a more powerful telescope and recognized the "ears" as rings.

Galileo didn't stop there. He further validated Copernican theory when he showed that like the Moon, Venus goes through a full set of phases, an impossibility if Venus and the Sun were orbiting a stationary Earth. He proved that the bright, cloudy patches of the Milky Way are composed of countless individual stars whose light has blurred together. These findings suggested that the universe is much bigger than many had expected. With each observation, Galileo added another nail to the coffin of geocentrism.

Another major shift in our cosmic perspective occurred the following century, when musician-turned-astronomer William Herschel constructed a 20-foot reflecting telescope. It had a much larger opening than the little telescope Galileo used for observing the moons of Jupiter, which meant it could collect a lot more light and was more sensitive to detail. In 1784 and 1785, he and his sister Caroline used it to count stars and construct a map of the Milky

---

* Simon Marius, a German astronomer, later gave the moons these names (at the suggestion of Kepler!). Marius independently observed the moons the same year Galileo did.

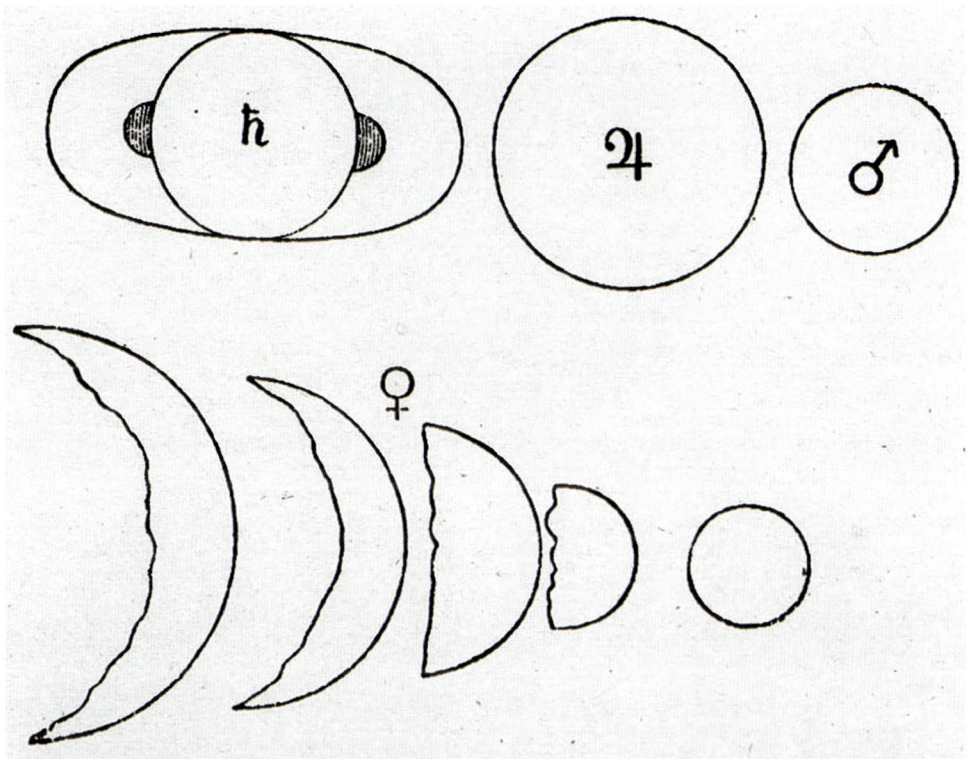

GALILEO'S OBSERVATIONS OF THE PLANETS | Galileo's drawings of Saturn, Jupiter, Mars, and the phases of Venus.

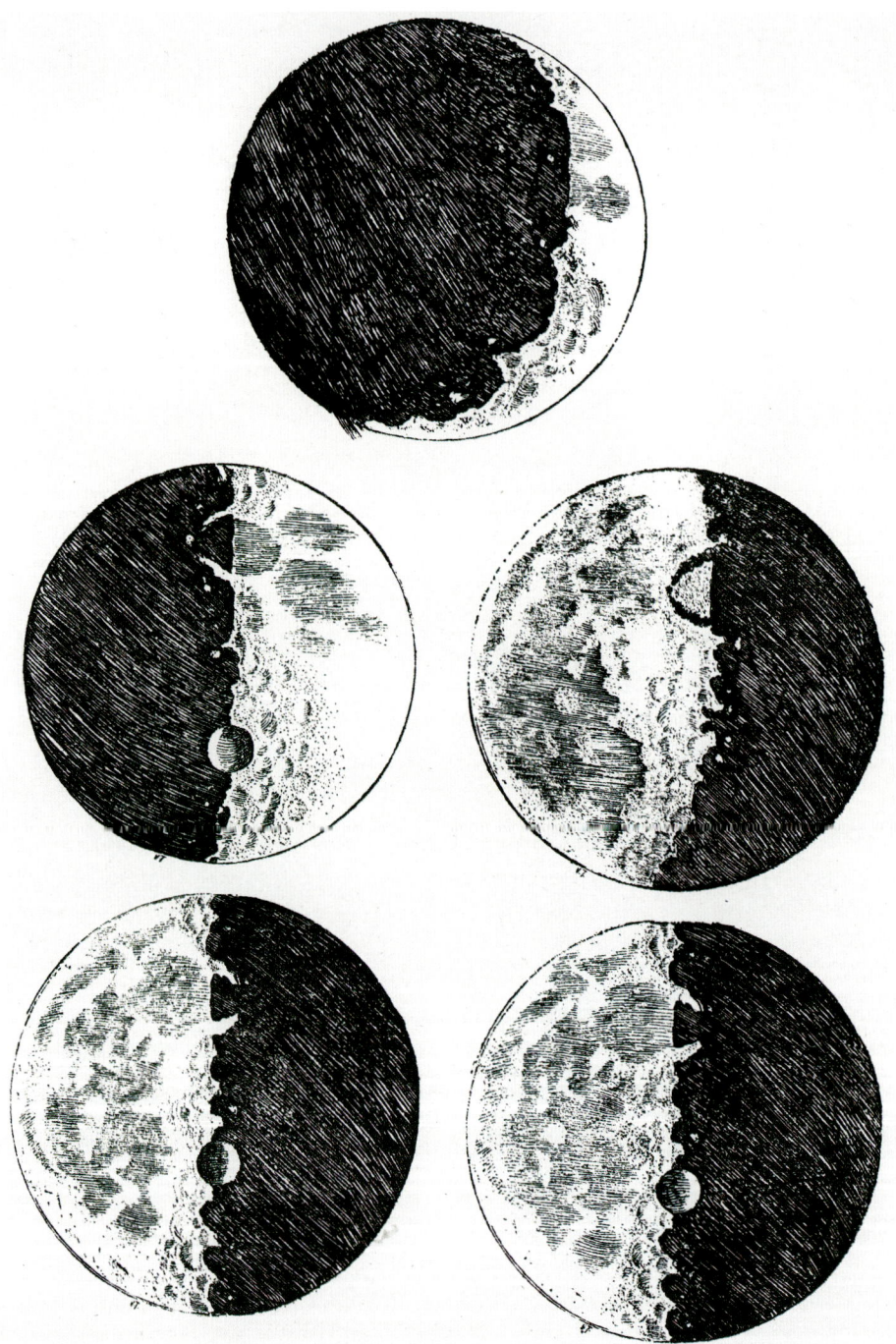

GALILEO'S DRAWINGS OF THE MOON | Galileo first observed the Moon through his telescope in 1609. These sketches and observations were published in his book *Sidereus Nuncius* ("*Starry Messenger*") the following year.

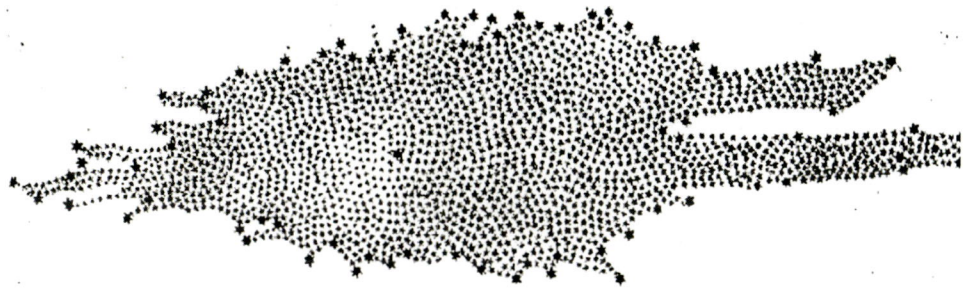

A VIEW OF THE MILKY WAY | William Herschel's 1785 star map of the Milky Way. The heavy asterisk near the center marks his presumed location of the Sun.

Way. Most of the stars they could see were arranged in a somewhat flat disk shape that circled the sky. Since the number of stars was roughly the same in all directions, they concluded that the Sun must be located near the center.

While the Herschels got the overall form of the Milky Way disk right—our Galaxy does, in fact, have a wheel-like shape that is much wider than it is thick—they got the Sun's location wrong. The Herschels didn't know that we live in a dusty Galaxy. Interstellar dust is concentrated in the disk, absorbing starlight and obscuring our view of distant stars. Because the Herschels had a limited perspective—they could see within a radius of only about 6,000 light-years—they concluded that our Solar System is located near the center of the Galaxy.

Herschel was humble about his findings and the assumptions that had led to them. He realized that his telescope probably wasn't resolving all of the stars in the Milky Way, that telescopes with larger apertures reveal stars invisible to smaller ones.* Even with his 40-foot telescope, built a few years later, Herschel still could not see to the farthest reaches of the Galaxy. A German immigrant to England who had had success as a professional musician before turning to astronomy, Herschel appreciated how "seeing is in some respect an art, which must be learnt."

---

* The *resolution* of a telescope refers to its ability to reveal detail, to create sharp images. Bigger telescopes collect more light than smaller ones and have higher resolution.

The art of learning to see nature leads to discovery. The process of science entails gradual revelation, a continuous peeling back of one after another of the universe's infinite, interconnected layers.

Astronomers peeled back another layer in the 1890s, when they discovered a type of *variable star* whose brightness changes with time. RR Lyrae variables are low-mass stars whose luminosity periodically goes up and down by as much as 10 to 20 percent over the course of a couple hours to a couple days. These variations occur because the stars pulsate in and out. An RR Lyrae gets brighter when its surface expands and dimmer when it contracts. When astronomers realized that the frequency with which the stars' brightness changes is predictable, dependent on their overall luminosity, they knew they'd received a gift from nature. They could use this predictable pattern to determine the *distances* to RR Lyraes.

This is what Harlow Shapley did in the early 1900s. He wanted to find out how globular clusters—dense groups of old stars containing tens to hundreds of thousands of members—are distributed in the Milky Way. It turns out that RR Lyrae variables are commonly located in globular clusters, so Shapley compared the actual luminosity of RR Lyraes to how bright they appeared on Earth in order to calculate their distances. When he discovered that globular clusters are scattered uniformly around a point in the Galactic Disk, he made the bold assumption that *this* point and not the Sun, as Herschel had thought, was the center of the Milky Way. He was right. Today we know that the Milky Way contains at least 150 globular clusters, mostly located in what we call the *halo* of the Galaxy.

The Solar System, on the other hand, lies in the outskirts of the Galaxy, about 27,000 light-years from the center. The brightest stars reside in the disk and are organized into spiral arms, while the oldest tend to be found in compact globular clusters distributed throughout the halo. The Sun, a middle-aged star, is located in the Orion Arm, a minor spiral arm between the larger Perseus and Sagittarius Arms. In 2020, the most detailed map of stars in the Milky Way to date was released by the Gaia mission, a project of the European Space Agency. Constructed from data gathered by the Gaia spacecraft over a six-year period, the map is like a digital patchwork quilt. It is woven with the distances, positions, brightnesses, and proper motions for 1.8 billion stars, revealing an exquisitely detailed view of the structure and dynamics of the Milky Way from our perspective in the Orion Arm. This map enables scientists to reconstruct the history of the Milky Way via a sort of galactic archaeology.

Imagine taking this image and wrapping it around yourself until one end touches the other; this will give you a 360-degree view of the Galaxy. The brightest region toward the center is a compact group of stars known as the galactic bulge. Most of these stars are about twice as old as the Sun and the other stars in our neighborhood. In the heart of the galactic bulge are more than 1,500 stars per cubic light-year—hundreds of thousands of times denser than the spacing of stars elsewhere in the Galaxy! We also see globular clusters sprinkled across the sky, as well as smaller *dwarf galaxies* orbiting the Milky Way.

Two of the closest dwarf galaxies appear prominently below the disk in the Gaia image—known as the Large and Small Magellanic Clouds, they are both visible from the Southern Hemisphere. They get their names from Ferdinand Magellan, the Portuguese "explorer," as he is commonly referred to in the West, though many consider him an agent of Spanish colonialism. Because the pair of galaxies can be seen with the naked eye, the indigenous people of South America certainly were aware of them long before Magellan came along in the 1500s. And the earliest written record of observations of the Large Magellanic Cloud was made hundreds of years before Magellan by the Persian astronomer Al-Sufi, who called it the White Bull.

In order for the light to shine so brightly, the darkness must be present.
— Francis Bacon —

As a child growing up in Washington, DC, Vera Rubin had always found it "more interesting to watch the stars than to sleep." Like Elizabeth Scott, she loved the stars. Years later, as a professor at Georgetown University in the early 1960s—like Harlow Shapley before her—Rubin asked a simple question:

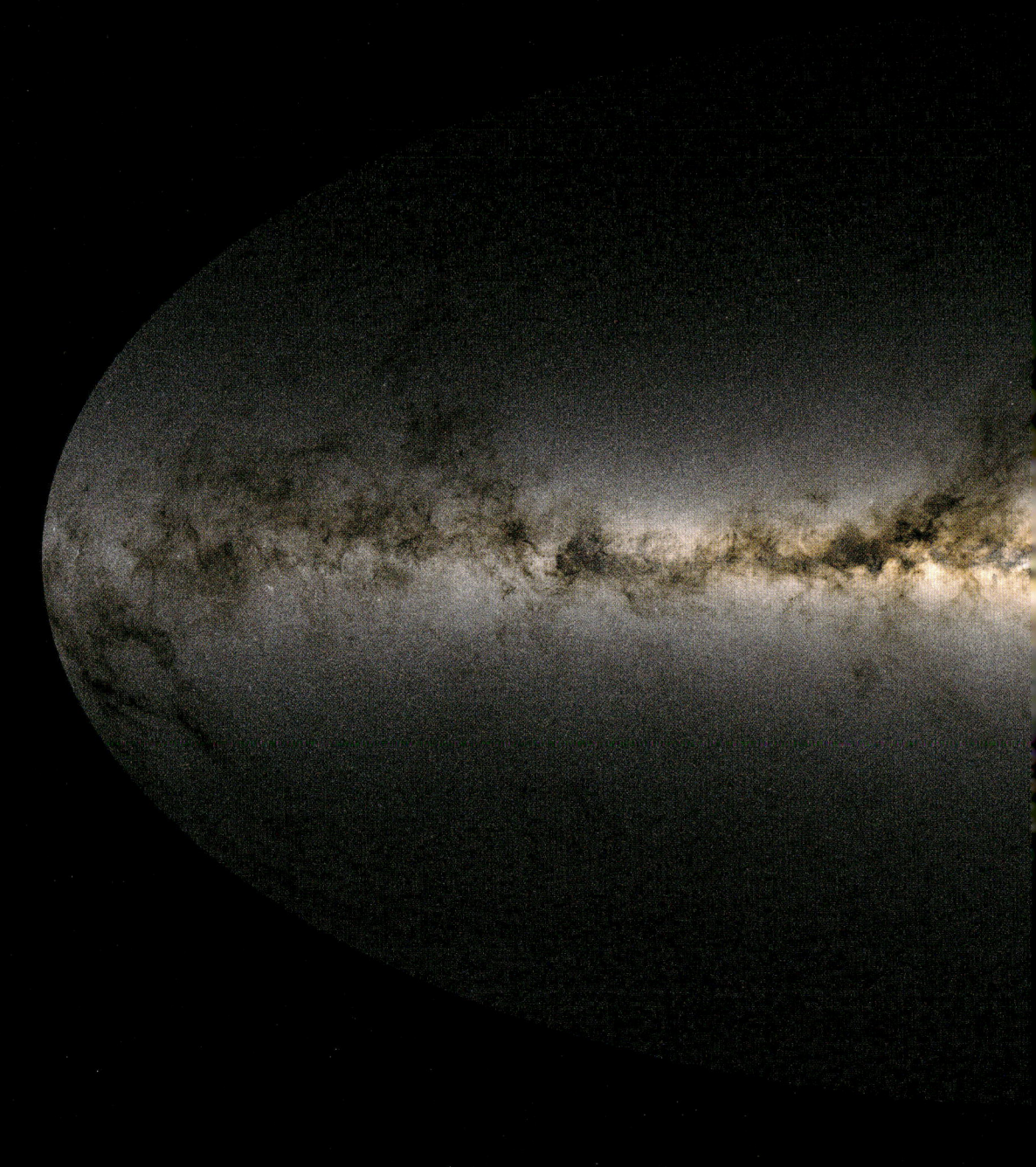

GAIA MAP OF THE MILKY WAY | Below the prominent disk of our Galaxy, just right of center, lie two of our closest galactic neighbors: the Large and Small Magellanic Clouds.

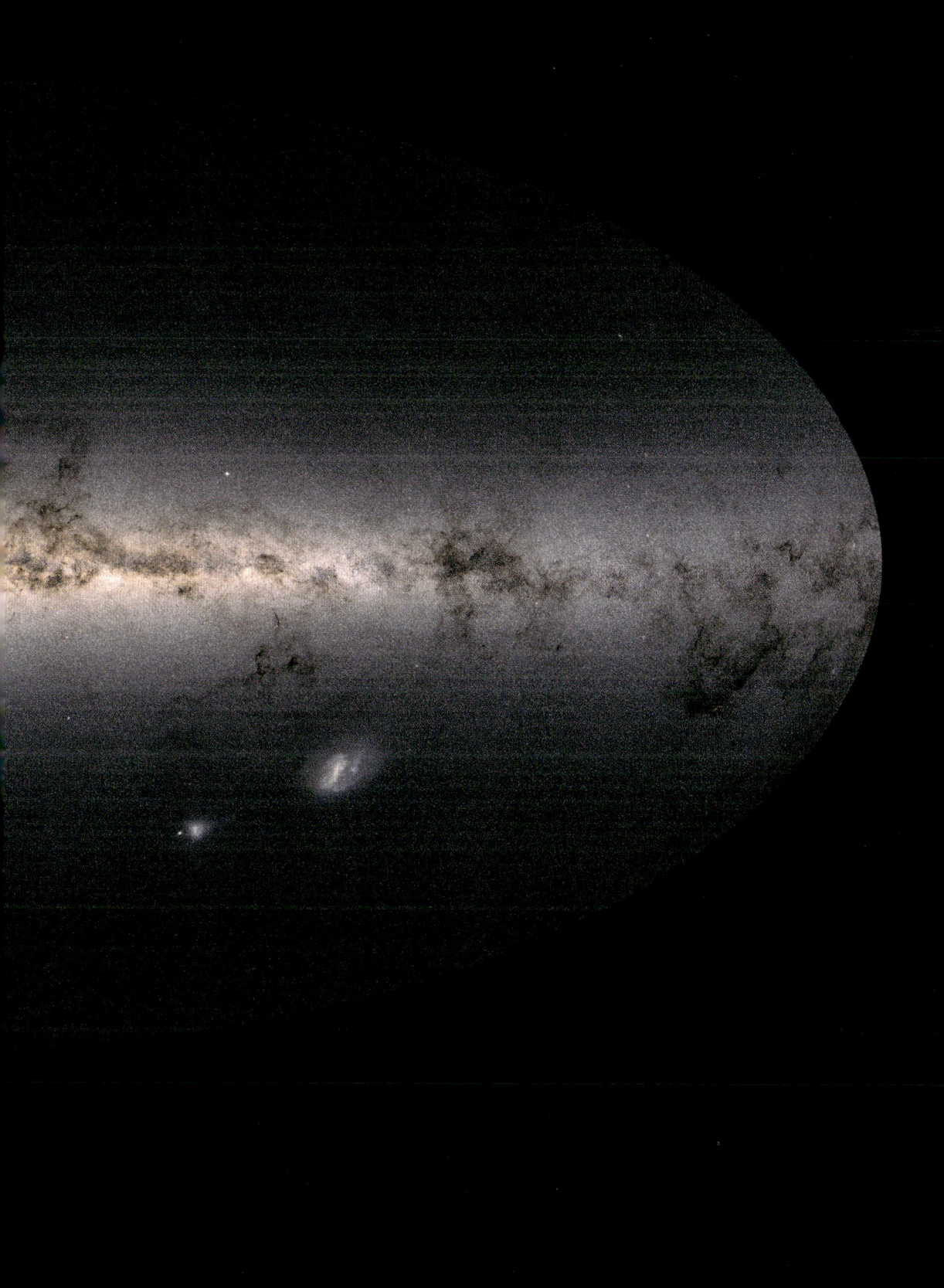

How do stars located *far* from the center of our Galaxy tend to move? The answer would forever alter our perspective on our place in the cosmos.

Most of the stars in the Milky Way, together with vast interstellar gas clouds, orbit the center of the Milky Way at hundreds of miles per second. Rubin was curious about how stars move in the distant outreaches of the Galaxy. She made a plot called a rotation curve, which shows how the speeds of stars change as a function of distance from the center of the Galaxy. A similar plot for the planets shows that as they get farther from the Sun, they slow down. Kepler demonstrated this in his third law, and Isaac Newton later showed why it happens: the Sun's gravitational pull gets weaker as distance increases. Might the same principle hold true for stars orbiting the center of a galaxy? Is it true that the farther they are from the center, the slower they go?

What Rubin and her students found is something quite different. Stars that are directly opposite the Galactic Center maintain roughly *the same* velocities. They don't slow down. As astronomers say, the rotation curve is *flat*. This observation raised puzzling questions. If we account for all the gravity present in the luminous matter in the Milky Way—this includes stars, planets, gas, and dust—the combined pull is *not* strong enough to keep the fast-moving stars in the outskirts from flying away. If we count up the entire amount of mass the Milky Way *must* have, as inferred from the motions of the stars, we get a far bigger number than can be explained by visible matter alone. In other words, *something* was clearly preventing stars from escaping the gravitational pull of the Milky Way. What could it be?

Rubin and her colleague Kent Ford began looking at our most massive galactic neighbor, Andromeda, to see if anything similar was going on there. From Kitt Peak National Observatory and Lowell Observatory in Arizona, they conducted observations to measure the rotation curve of Andromeda. This time, rather than using stars, they measured the light from nebulae known as *H II regions* at a range of distances from the galaxy center.* On

---

* "H II" is pronounced "H-*two*" and refers to ionized hydrogen. H II regions are associated with massive stars, whose powerful radiation is capable of stripping neutral hydrogen atoms of their electrons.

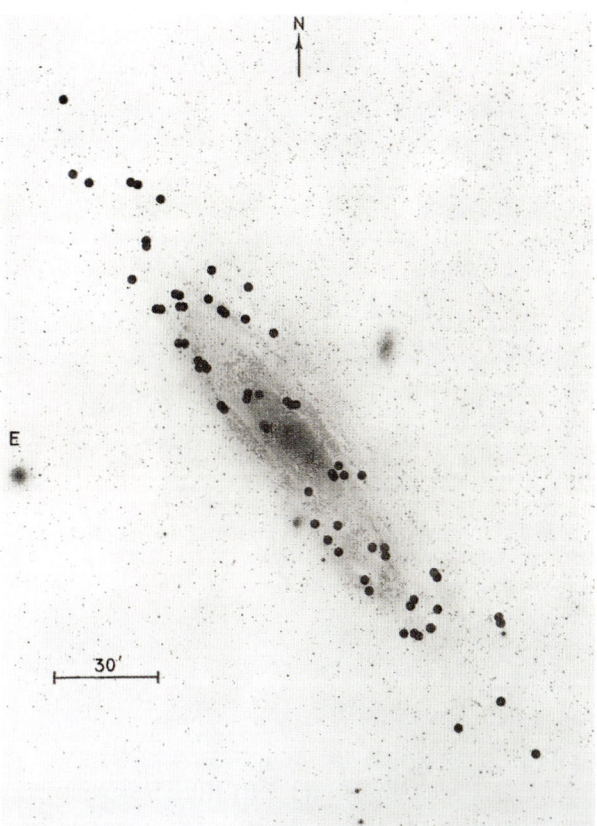

THE ANDROMEDA GALAXY | Vera Rubin's and Kent Ford's finding chart of H ɪɪ regions in Andromeda. *From Vera C. Rubin and W. Kent Ford, Jr., "Rotation of the Andromeda Nebula from a Spectroscopic Survey of Emission Regions," Astrophysical Journal 159 (February 1970): 379. © AAS. Reproduced with permission.*

the first night, as Rubin shifted between eating ice cream cones and developing photographic plates, they made a discovery. Andromeda also has a flat rotation curve. As Rubin recalls in an autobiographical narrative, "I remember thinking that there must be some mechanism for speeding up stars that moved too slowly or slowing down stars that moved too fast."

Today the accepted explanation for flat rotation curves is that an additional source of gravity must be distributed throughout galaxies and that it extends far past the outermost stellar orbits. *Dark matter* is the invisible "glue" that keeps these fast-moving stars from escaping.

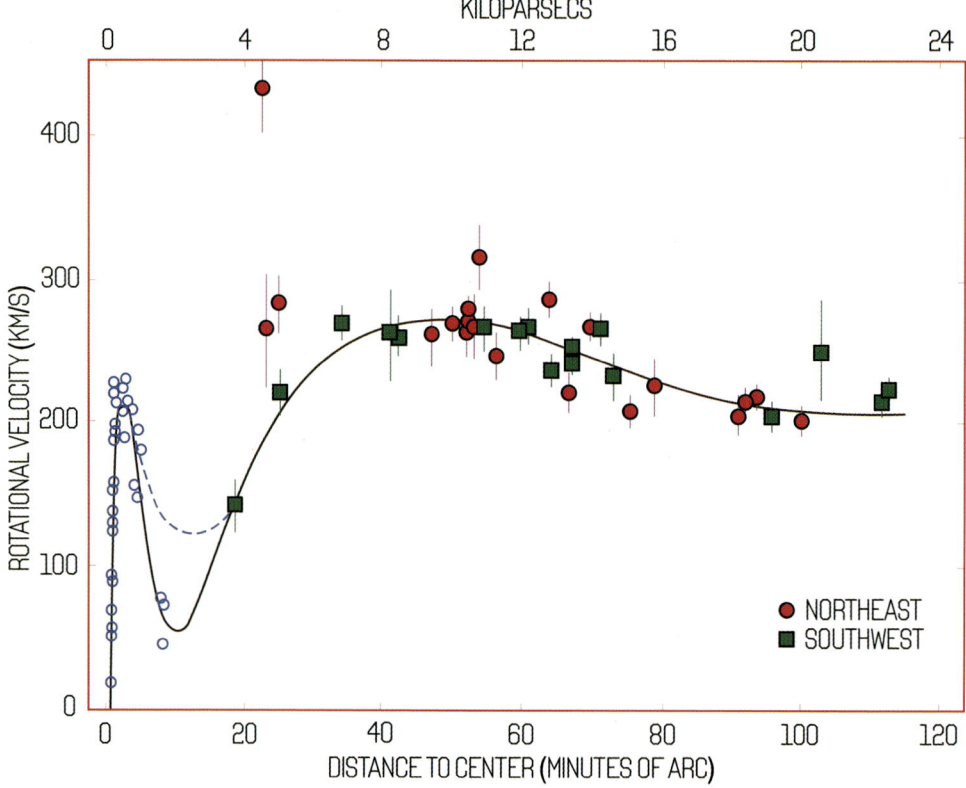

ROTATION CURVE OF ANDROMEDA | A color reproduction of Rubin and Ford's original plot from their 1970 paper. Rubin and Ford observed H II regions associated with massive stars in Andromeda and measured their speeds. Each point on this plot represents the speed of a region and its distance to the center of the galaxy. Beyond roughly 8 kiloparsecs (a little more than 26,000 light-years), the rotation curve "flattens out," an indication of the presence of dark matter in Andromeda. *ADAPTED FROM RUBIN AND FORD, "ROTATION OF THE ANDROMEDA NEBULA FROM A SPECTROSCOPIC SURVEY OF EMISSION REGIONS."*

Dark matter is a form of nonluminous matter making up the bulk of the mass of the Milky Way and other galaxies. We call it "dark" because it does not emit or interact with radiation from any part of the electromagnetic spectrum. Though we can't see it, we know it's there because its gravitational influence on radiant celestial bodies is unmistakable. In the Milky Way, dark matter extends in a spherical halo some 200,000 light-years from the center, spanning a diameter at least twice as far as the luminous matter. There is about 20 times as much dark matter as "normal" visible matter detected with telescopes.

Fritz Zwicky was the first to predict the existence of invisible matter, some 30 years before Rubin and Ford's work. Observing the speeds of galaxies in the Coma Cluster, he realized that the gravity from visible matter in the cluster was not strong enough to keep it from dispersing. Galaxy clusters are the most massive entities in the universe, containing anywhere from 100 trillion to a quadrillion solar masses of material.* We now know that the great bulk of this mass takes the form of dark matter; stars and gas represent only the tip of the cosmic iceberg.

In 2022, NASA revealed an image of the galaxy cluster SMACS 0723, taken by the James Webb Space Telescope. The picture is a composite of images at four different near-infrared wavelengths that are translated into visible color. The cluster contains thousands of individual galaxies of different shapes, sizes, and masses. Several of the galaxy images are stretched out into arcs that collectively form a sort of ring on the sky. If you look carefully, you can see that some stretched-out galaxies are duplicated. This image shows SMACS 0723 as it was 4.6 billion years ago. But these ancient galaxies are much farther away—the light from the most distant of them took more than 13 billion years to reach us. They actually lie *behind* the cluster. Their images are warped and multiplied due to *gravitational lensing*, a type of distortion caused by gravity. Most of the matter causing gravitational lensing in SMACS 0723 is dark.

---

* A hundred trillion (10^{14}) is the number 1 followed by 14 zeros. A quadrillion (10^{15}) is the number 1 followed by 15 zeros: 1,000,000,000,000,000.

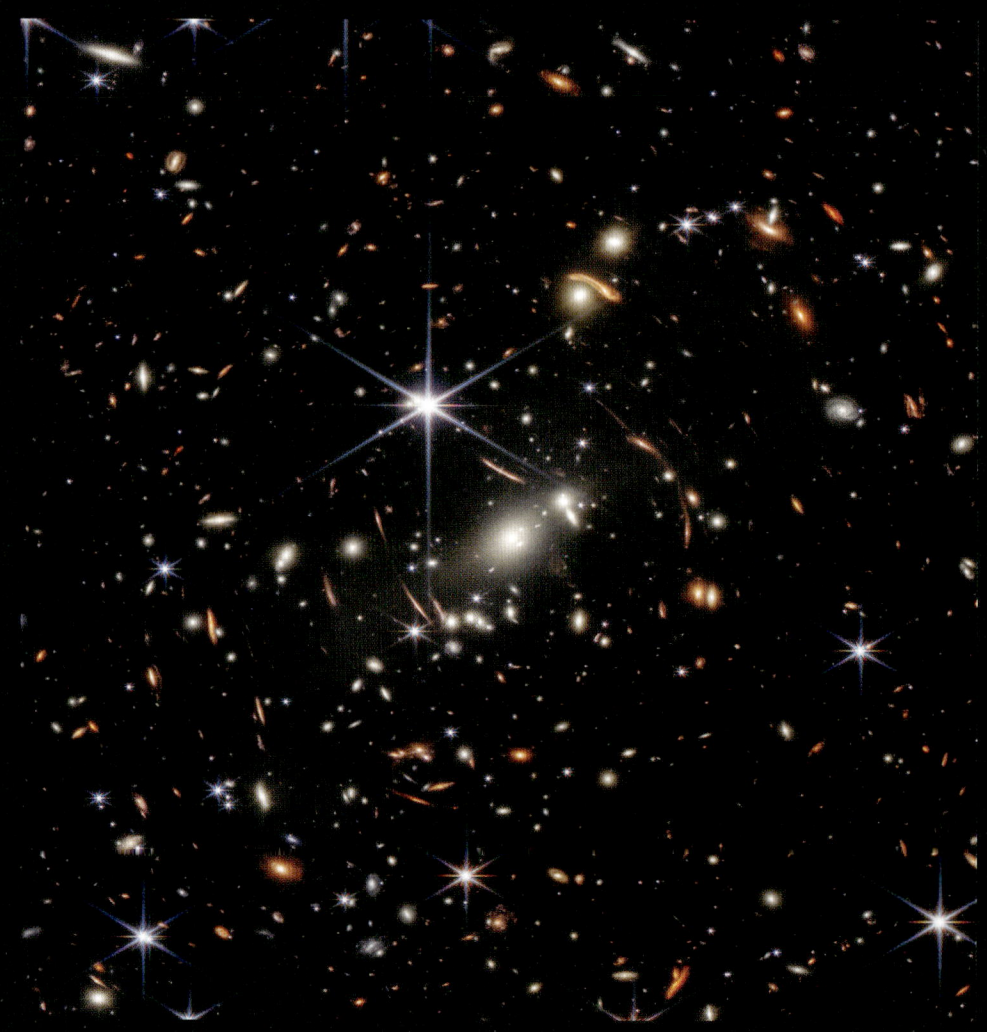

SMACS 0723 | Infrared image of the SMACS 0723 galaxy cluster taken by the James Webb Space Telescope.

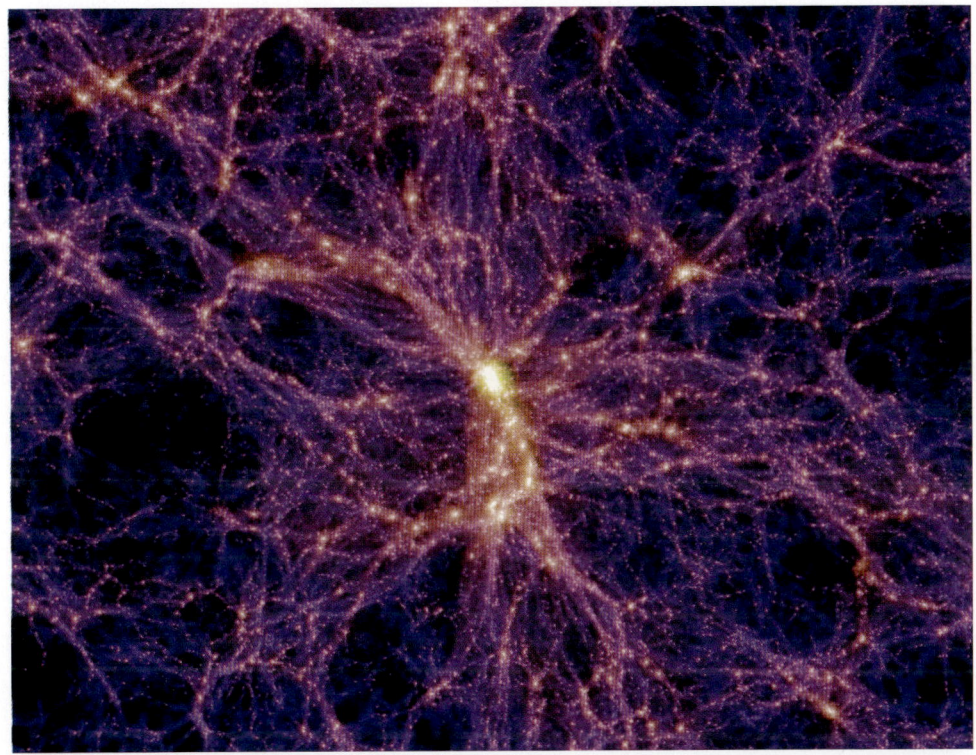

SIMULATING DARK MATTER | A computer simulation by the Millennium Simulation Project visualizing the distribution of dark matter on very large scales in the universe. The number 31.25 Mpc/h corresponds to a length of about 145 million light-years. Clusters of galaxies form where concentrations of dark matter are the highest, at junctions in the vast cosmic web.

In the decades since Rubin and Ford's discovery, many experiments have been conducted to directly detect dark matter particles. Maybe dark matter consists of exotic subatomic particles never before detected on Earth. If so, what is the mass of a fundamental unit of dark matter? How might these hypothetical particles be distributed throughout space? The Vera C. Rubin Observatory, currently under construction in Chile, will house a telescope that astronomers will use to answer these questions by measuring gravitational lensing. Like a glass lens, gravity can distort images. Although dark matter cannot be detected directly, astronomers hope to infer its intrinsic properties by observing how the light from distant galaxies is distorted when it passes through concentrations of this invisible material.

For now, the only thing astronomers know for certain is what dark matter is *not*. Indirect evidence shows that dark matter can't be made up of lots of small, very-low-mass bodies, such as brown dwarfs, which escape detection due to their dimness. Neither could dark matter be comprised of lots of black holes. If it were, the Galaxy would be filled with much more x-ray emission—generated when black holes rip away material from nearby stars and gas clouds—than we actually observe. It is now believed that 85 percent of all matter in the universe is dark, of a nature entirely unlike the "ordinary" matter that we and everything we are familiar with are made of. Research from the Millennium Simulation Project and other groups indicates that dark matter pervades intergalactic space like a cosmic spiderweb, and clusters of galaxies assemble where the concentration of dark matter is highest. But still, we have no idea what dark matter is. Once again, human beings are displaced from the center.

"God's Hand Staid the Stars"

Without humility, we limit ourselves to narrow questions and a narrow way of seeing the world. My experience in science has taught me that for every question answered, a myriad of new ones emerges. Knowing that, around any corner, surprising new vistas may open up that radically alter our perspective about humanity's place in the universe, the stargazer would do well to remain humble.

Art can help us cultivate humility if we are receptive to it. Art invites us to share the perspective of another, lifts us out of the prison of our egoism, and expands our capacity to feel, think, and connect with life beyond ourselves. No artform illustrates the virtue of humility more than quilt making. An art with roots in many different cultures, quilting can be practiced by those with the meagerest means. The quilter labors for days, meticulously piecing together diverse scraps of fabric and then stitching the layers together into a thing of beauty. Sometimes the work is done by a collective. The finished product will come into intimate contact with its user; it will be used, washed, exposed to the elements, and become worn and faded. A quilt can be made entirely by hand, using unassuming fragments from secondhand clothes or even pieces from old quilts. To be wrapped in such a quilt is to be wrapped in a tangible expression of humility.

Quilts made by blacks in antebellum America often served many purposes. In addition to providing warmth, a quilt could tell a story; it could be a visible record of family history. An unpretentious blanket might be woven with secret messages that broadcast where a freedom-bound slave could go to find safe shelter. The bold geometric patterns, knots, stitching, and color choices all worked together to signify meaning to those initiated in a particular tradition. For enslaved blacks who came from cultures with rich oral traditions, the signs and symbols in a quilt were the visual equivalent of speech.

Some symbols were brought from Africa, from people with sophisticated traditions in textile art, such as the Igbo people of Nigeria; the Asante of Ghana; the Dahomey of Benin, famous for their appliqué banners; and the Kuba of the Kongo, celebrated for their elaborately embroidered raffia cloth. Some of these ancestral symbols retained their original meaning among enslaved blacks. The Kongo cosmogram, for instance, a cross symbolizing the intersection between heaven and Earth, the living and the dead, appears in the art of enslaved blacks throughout the Americas, including rituals like the ring shout. Newer symbols were also incorporated into quilts and could take on an entirely different meaning, one that could be highly particular to the family or community to which the quilter belonged.

Coded language was critical to survival and resistance during slavery. Barred from literacy and open communication, enslaved people developed sophisticated ways of sharing knowledge and communicating in secret. Quilts were just one of the instruments that could encode hidden meaning. They often worked in concert with songs. The lyrics of Negro spirituals were highly metaphorical and could be sung in the presence of unsuspecting whites who dismissed them as the childlike intonations of a simpleminded people.

A quilt might be one element in a complex system of coded communication used on the Underground Railroad. Through the nuanced selection of abstract, geometric patterns and color combinations, a quilter could hide in plain view a sign indicating that a house was a safe "station" or instructions on how to prepare for escape. To the unsuspecting passerby, the colorful quilt hanging on the clothesline along with the rest of the wash wouldn't raise suspicion. But the quilter has used blocks in specific patterns—such as the monkey wrench or flying geese—to signal to a group of slaves that it was time to prepare for embarking on their flight north by gathering the tools and provisions needed for their journey. In their book *Hidden in Plain View*, Jacqueline Tobin and Raymond Dobard explain how quilts also served as maps on the Underground Railroad: "A clever quilter would be able to specify one direction by the use of fabric pieces. For example, if a quilter wanted to indicate a northern direction, she/he would simply make one set of triangles distinct from the others."

Enslaved quilters and their descendants were not content to simply map Earth; they also charted the cosmos. The importance of the North Star in slave culture—as a navigation guide and a symbol of freedom—is well known. The famous song "Follow the Drinking Gourd" alludes to it, as do quilt blocks such as the eight-point star. But the North Star was not the only celestial body that inspired these generations of black quilters. Born into slavery in rural Georgia, Harriet Powers would become known for creating phenomenal quilts that illustrated stories from the Bible. Her striking appliqué figures, reminiscent of those used in Dahomey banners, depict not only Bible scenes but also

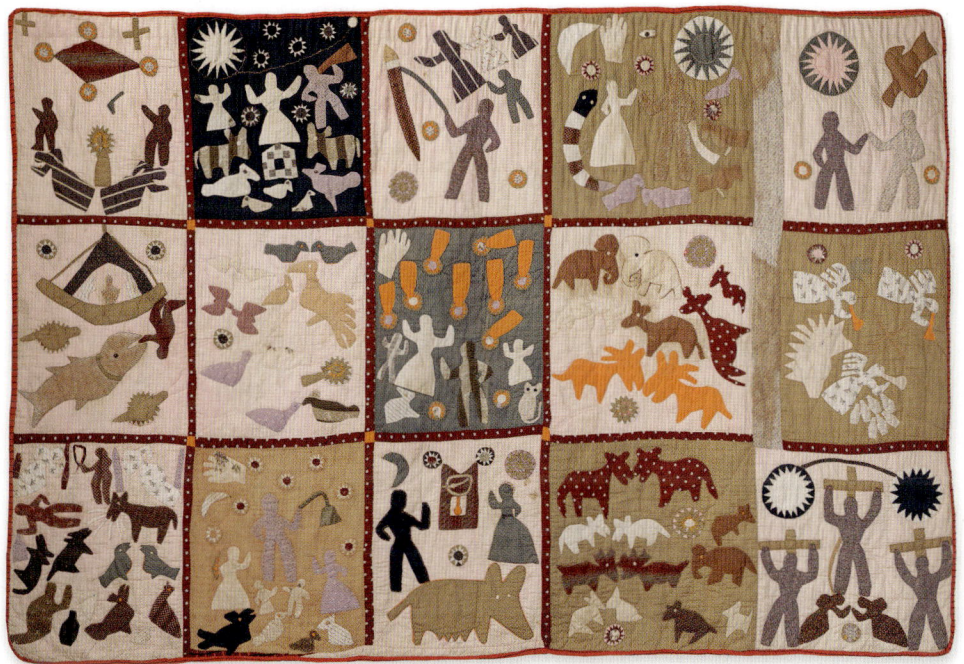

PICTORIAL QUILT | Harriet Powers, 1895, cotton textiles, 68⁹⁄₁₀ × 105 in (175 × 266.7 cm). The Leonid meteor storm of 1833 is depicted in the third block of the second row. "God's hand" appears in this and two other blocks. *Photograph © Museum of Fine Arts, Boston.*

astronomical events. In one of her quilts, Powers depicted the Leonid meteor storm of 1833. The Leonids are an annual celestial event active in November, when Earth passes through the trail of debris left by the comet known as Tempel-Tuttle. Usually peaking in mid-November, these showers may give a view of about ten meteors per hour. Every 33 years—the orbital period of Tempel-Tuttle around the Sun—the event goes from shower to *storm*, and *thousands* of meteors per hour may be observed radiating across the sky.

In 1898, Powers's *Pictorial Quilt* was displayed at Atlanta University accompanied by her description of each block, including the one portraying the Leonids: "The falling of the stars on Nov. 13, 1833. The people were frightened and thought that the end had come. God's hand staid the stars. The varmints rushed out of their beds." I find it fascinating that the Leonid storm occurred four years *before* the artist was born, in 1837. Stories that Powers heard about the celestial event clearly made a strong impact on her, filling her with a sense of wonder.

The generations of black quilters born after slavery integrated elements of their ancestors' style and technique into their art, building upon that foundation while giving an entirely new direction to the artform. Women like Elizabeth Talford Scott, Nettie Lee Young, Laverne Brackens, and Rosie Lee Tompkins grew up picking cotton and were inducted into the art of quilt making through their mothers, aunties, or other women in the community.

When I first encountered Tompkins's work, I felt stunned by the power of her art. I felt I was in the presence of a colorist of the same caliber as Romare Bearden or van Gogh, a storyteller as virtuosic and nuanced as Langston Hughes or Toni Morrison. Tompkins's quilts belong to the African American tradition of improvisational quilt making, and she incorporated all sorts of unexpected materials into her work—lush velvets, faux furs, embroidered cloth, secondhand clothes, T-shirts, men's silk ties, sequined fabric, rhinestones. In a quilt from 1996, she arranged pieces of an American flag, dish towels, a calendar, and a textile printed with the face of a white Jesus into an irresistible composition. Improvisation is a defining characteristic of African American art and culture, showing up especially in black music—like jazz and hip-hop—but also in visual art. Tompkins didn't use patterns, but there's nothing random about her work; her masterful improvisation fills her art with little surprises and conveys a quality of immediacy, of freedom and controlled power.

Rosie Lee Tompkins (a pseudonym) was born Effie Mae Martin and moved west as part of the postwar Great Migration. She died in 2006, having received some well-deserved acclaim for her prolific body of work, much of which is now housed at the Berkeley Art Museum in California. Like Tompkins, but in her own inventive way, the contemporary quilt maker Bisa Butler creates work rich with stories and linkages to African and black American culture. Her life-sized quilted portraits, with their vivid, kaleidoscopic color, remind me of the collages of Romare Bearden; at first look they appear more like paintings than fabric art. Like Tompkins, Butler has revolutionized a medium.

She often incorporates African textiles such as Kente cloth into her palette, thus weaving together traditions from two continents. Her choice to depict people of African descent in this way reflects her training at Howard University, where she learned from professors who were part of AfriCOBRA, an art collective that emerged from the Black Power movement in 1968. The African Commune of Bad Relevant Artists aimed to create images that empower members of the African diaspora, promote solidarity among them, and reflect the community. Their commitment to expressing the depth and continuity of African and black American culture feels palpable in Butler's work. "I think about myself being like one in succession of this line of quilters," she said in an interview. "And that's one thing that I really love about quilting—that it grounds you in this history of women creating."

By "grounding" herself in history—like Harriet Powers, Elizabeth Talford Scott, and Rosie Lee Tompkins before her—Bisa Butler has attached herself and her community to a larger grand narrative. In parallel, Ptolemy, Copernicus, Kepler, Galileo, and Rubin belonged to a tradition of scientists who sought humanity's connection to the stars—scientists who, by drawing from the lessons of history, attempted to quilt together our cosmic and terrestrial stories. By attaching ourselves to something bigger than ourselves—an act that requires a posture of humility—we heighten our awareness of the interconnectedness and interdependence of all.

CHAPTER TEN

FAITH

*Faith indeed tells what the senses do not tell, but not the contrary of what
they see. It is above them and not contrary to them.*
— Blaise Pascal —

*Faith is the strength by which a shattered world shall emerge
into the light.*
— Helen Keller —

In 1965, a single mother of two young boys began writing a novel about a little black girl who prayed to God to give her blue eyes. God didn't. But an unscrupulous neighbor made Pecola Breedlove believe her prayer had been granted, upon which the abused, neglected child began to lose her mind. The tragic story shows how racism manifests in the psyche of the most vulnerable, defenseless members of American society, from the perspective of those preyed upon. Toni Morrison explained why she wrote *The Bluest Eye*: "I wrote the first book because I wanted to read it. I thought that kind of book, with that subject—those most vulnerable, most undescribed, not taken seriously little black girls—had never existed seriously in literature. No one had ever written about them except as props."

The Bluest Eye was the first piece in a body of work that would revolutionize Western literature and transform how people across the globe understand and relate to the black American experience.

Faith is an essential ingredient for creation. It takes many forms, depending on the creator and their vision. For Toni Morrison, faith took the form of writing novels; the vision involved creating a literature that did justice to the stories of a people who have been subjugated and shamelessly exploited in literature and life, yet whose struggle have produced great beauty in the world. Faith is an indispensable part of being an artist, a scientist, or any creative person who attempts to inject a new vision into the world.

In a 2013 speech at Vanderbilt University, Morrison said, "I am a writer and my faith in the world of art is intense but not irrational or naïve. Art invites us to take the journey beyond price, beyond costs into bearing witness to the world as it is and as it should be. Art invites us to know beauty and to solicit it from even the most tragic of circumstances." Such faith requires vulnerability and the courage to do something never done before, to step out into the unknown where failure is an entirely possible outcome. It requires ambition—to grow, to evolve—and a recognition that we are part of a world itself constantly changing, as is our understanding of it.

Island Universes

We all have faith in something. Scientists are no exception. Well into the 20th century, most American and European scientists believed the universe is static. An unchanging cosmos was an article of faith as much as Newton's law of gravity. But evidence to the contrary had steadily been mounting, and it was becoming clear that a choice needed to be made: keep trying to conform nature to old beliefs or reexamine these beliefs in light of new facts.

One fact that astronomers and physicists of the time couldn't ignore concerned the puzzling orbit of Mercury. By the late 1800s, a few centuries'

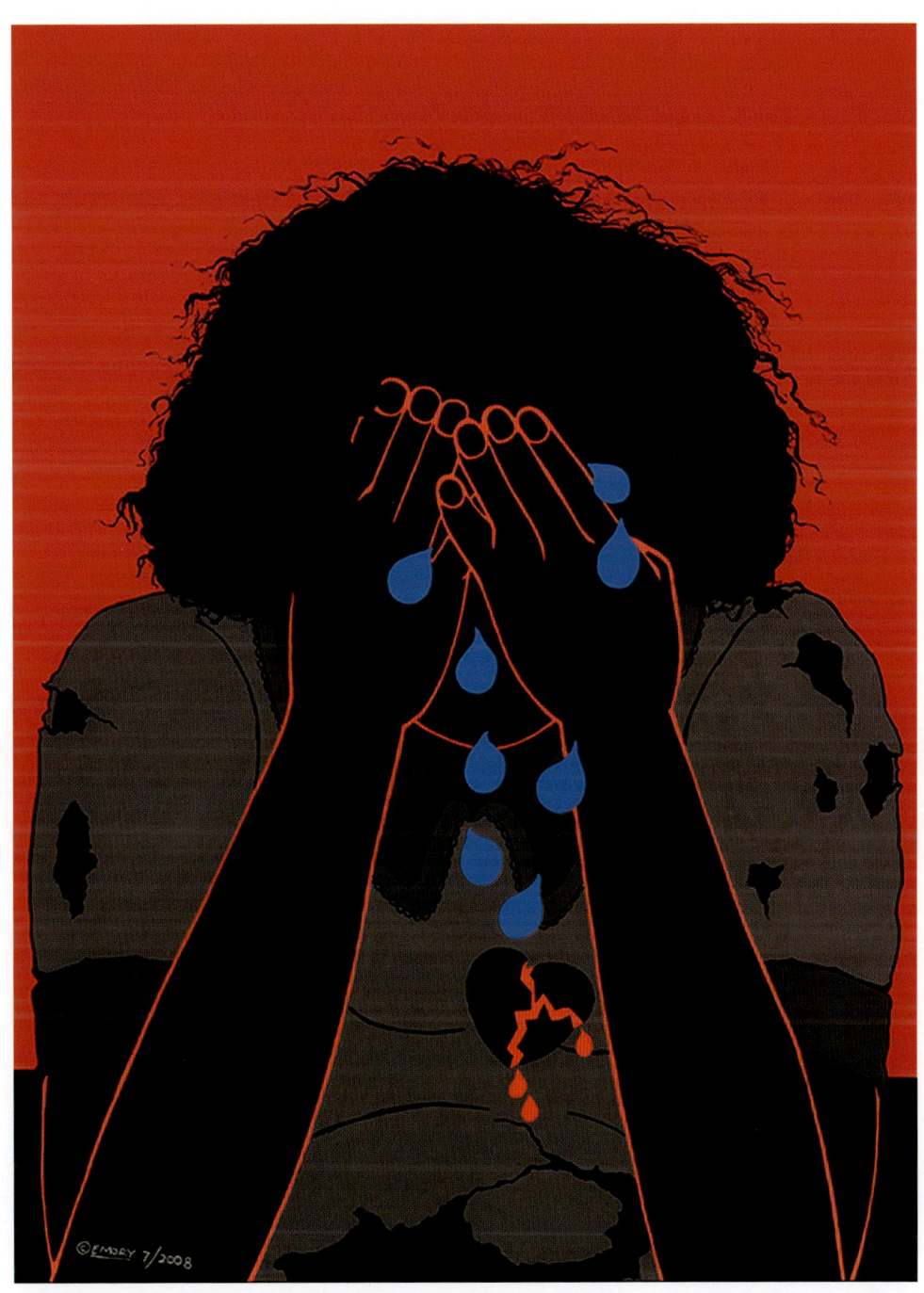

THE ASSASSINATION OF PECOLA BREEDLOVE (THE BLUEST EYE) | Emory Douglas, 2008, acrylic paint on masonite board, 18 × 24 in (45.7 × 61 cm).

worth of observations showed that Mercury's orbit was shifting around by an amount that couldn't be explained by Newton's theory of gravity. To resolve this discrepancy between observation and theory, scientists invoked everything from the presence of a hidden inner planet that could be perturbing Mercury's orbit to the existence of forces other than gravity that might be acting on the planet. A few brave souls attempted instead to reexamine Newton's law of gravity.

Einstein was among those who pursued the latter route, and ultimately, he found the solution with his general theory of relativity, which explains how gravity emerges from the properties of nature. Earlier he had developed the theory of *special* relativity, discussed in Chapter 7. It explains how mass, time, and space change as moving objects approach the speed of light. Published in 1905, the theory makes use of a powerful idea that originated with Einstein's former teacher Hermann Minkowski, that space and time are malleable and should be considered part of the same essential "fabric" of nature.

Over the next decade, Einstein developed a more *general* theory of relativity, which considers what happens when objects are no longer moving at a constant velocity, as in special relativity, but accelerating with respect to each other. This required a change in the assumptions about the fabric of spacetime. In special relativity, spacetime is taken to be flat. But general relativity introduced a new idea: spacetime may be *curved*.

To return to an earlier analogy—imagine a taut piece of fabric, like a bedsheet stretched tight over an infinitely large area. If you sent a small marble rolling on it, absent any other forces, the marble would roll along forever in a straight line at the same speed. Now imagine putting down a bowling ball. The sheet dips down; it curves. If our marble comes rolling by, it will fall toward the bowling ball, picking up speed as it does.

In this analogy, the fabric is spacetime, the bowling ball could be the Sun—or matter of any kind—and the marble could be a planet. The presence of mass and energy is what causes the curvature of spacetime. This is the key idea of general relativity, which describes the relationship between the geometry of spacetime and the distribution of matter in it with a set of equations. The equations describe how gravity is embedded into the very geometry of spacetime.

General relativity marked a great leap forward from classical Newtonian physics, which quantifies the force of gravity but does not explain its essential nature. Newton's law of gravity that we learn about in high school physics says that the strength of gravitational attraction between two bodies depends on how massive they are and the distance between them. This way of thinking about gravity is good enough when the strength of gravity is relatively weak. But close to a very strong source of gravity, like the Sun, Newton's theory breaks down, as it also does when objects are moving very fast, approaching the speed of light. General relativity is needed to explain what's going on in these cases.

Einstein used his theory to correctly predict how Mercury's orbit should shift, giving him and other proponents of general relativity confidence that they were on the right track. But more evidence was needed to confirm its principles and convince skeptics.

Four years after its publication in 1915, general relativity would receive further validation. As the total solar eclipse of 1919 approached, the astronomer Arthur Eddington jumped at the opportunity to test general relativity by observing whether light—which has no mass—responds to the bending of spacetime just as massive objects do. If Einstein was right, we should be able to observe the bending of light by looking at stars appearing near the Sun. When light from a distant star grazes the surface of the Sun, the theory predicts that it should bend, as it follows the curvature of space produced by the Sun's enormous mass. From our perspective on Earth, the star's location will seem shifted away from its normal position by a tiny amount. The eclipse of 1919 offered an opportunity to confirm this phenomenon, because when the Moon obscures the Sun, it is possible to observe the apparent shift in the position of stars that are normally outshined by the Sun.

Arthur Eddington traveled to Principe, an island off the west coast of Africa, and sent other members of his team to Sobral, Brazil, to measure the deflection of starlight by the Sun and found that the angle by which it was deflected (1.75 arcseconds) agreed with Einstein's prediction.* The results of this effort made headline news and propelled Einstein into the international spotlight. His theory had truly surpassed Newton's conception of gravity.

Now that Einstein understood the workings of general relativity in the Solar System, he was eager to apply it to the entire universe. He pursued this goal while bearing in mind two assumptions common at the time: the universe is both static and infinite. The idea of a static universe makes intuitive sense in light of centuries of observations, which seem to indicate the generally fixed position of the stars. That the universe could be infinite in time, with no beginning or end, also seems plausible. But a nagging paradox quickly arises: If the universe is infinitely old and has an infinite number of stars, why is the night sky dark? In this hypothetical universe uniformly distributed with stars, you would expect to find a star along every line of sight, and the sky should always be ablaze with light.

To get around these problems, Einstein had to think creatively about the geometry of spacetime. He had already established that it was less like an infinitely broad sheet and more like the surface of a huge sphere curving back on itself. In such a "closed" universe, spacetime could have the remarkable property of being boundless yet finite. This assumption enabled Einstein to continue to work with the idea that the universe was stationary.

But there was another problem. What prevented the universe from collapsing in on itself due to the combined gravity of all the matter in it? To address this question, Einstein had to introduce a new term to his equations, one that delivered a "push" to balance the gravitational "pull" of matter. This force of repulsion came to be called the *cosmological constant*, which some considered a sort of fudge factor that ruined the simple elegance of Einstein's original equations. But it was mathematically necessary to describe a universe that, he believed, was static.

---

* One second of arc is 1/3,600 of a degree. A shift of 1.75 arcseconds is a tiny amount. By comparison, the angular size of the Sun is about 1,800 arcseconds.

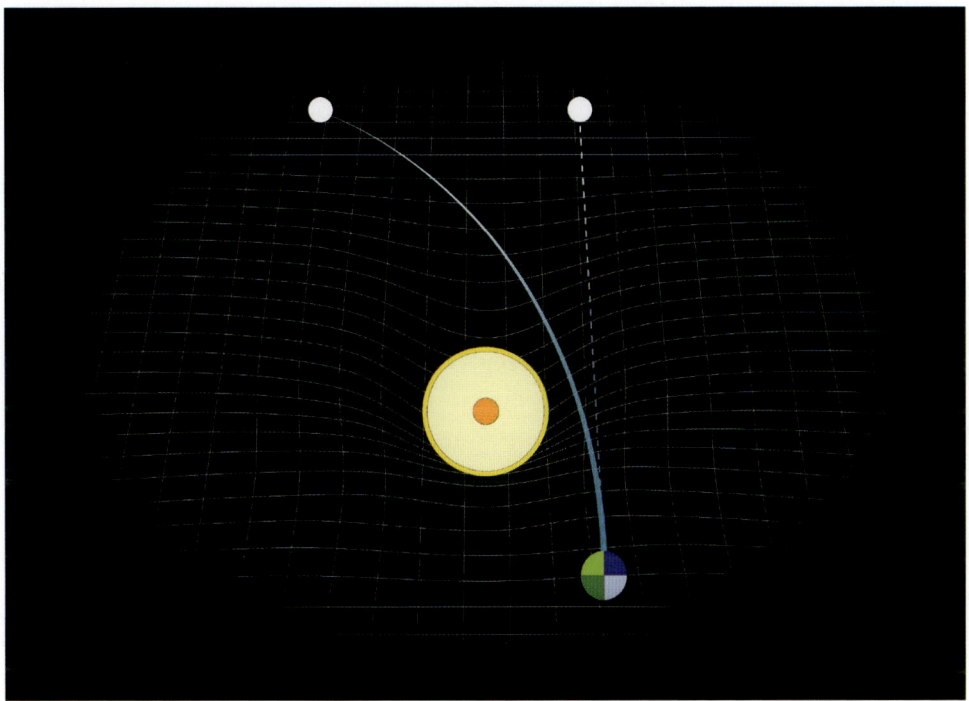

BENDING OF STARLIGHT NEAR THE SUN | During a total solar eclipse, the apparent tiny shift of a star from its true location in the sky can be measured because the Moon (not shown here) blocks the Sun's glare.

In 1922, five years after Einstein published the results of this model, a young Russian mathematician and physicist named Alexander Friedmann broke with the assumption that the universe was stationary. Friedmann proposed that Einstein's equations be considered in light of an *evolving* universe, and he came up with solutions that tied the cosmological constant to the structure and evolution and the cosmos. Einstein disapproved, publicly criticizing Friedmann for a mistake in his calculations. He quickly withdrew his censure when he discovered the mistake was his own (though an unpublished version of his retraction indicated that he still considered Friedmann's time-varying cosmology untenable).

The scientific community was slow to give heed to Friedmann's model and to a similar proposition put forth by physicist Georges Lemaître a few years later; perhaps this was because Friedmann was not widely known outside Russia and Lemaître's work was published in an obscure Belgian journal. Scientific progress takes time because knowledge is often transmitted inefficiently, and physical evidence supporting a theory is revealed only gradually, as new experiments are imagined and the technology required to conduct them becomes available. At the same time, scientific progress is often slowed because people are reluctant to break from convention for fear of censure and humiliation if proved wrong.

Meanwhile, during the period following World War I, observations of a mysterious kind of celestial object were forcing scientists to reevaluate their beliefs about the size and structure of the universe. As far back as the 18th century, astronomers had been aware of the existence of so-called "spiral nebulae," blurry patches of light in the sky noticeably different from normal stars. Scientists and philosophers debated what they might be. Were they groupings of many stars, so far away that their light was smeared out, or were they celestial bodies of a completely different kind? Were they located inside the Milky Way? Or were these fuzzy nebulae distinct "island universes," as the philosopher Immanuel Kant suggested, lying far beyond our Galaxy?

By the early 1900s, advances in telescope technology enabled astronomers to not only see more detail in the structure of spiral nebulae, but to obtain measurements of their motions. When a celestial body, such as a star or nebula, is moving toward or away from Earth, its light waves will shift by an amount corresponding to its speed. If the object is moving toward us, from our perspective the light it emits will be compressed toward shorter wavelengths—we say the light is *blueshifted*. If on the other hand the object is receding, its light will be elongated or *redshifted*. And the faster the motion toward or away, the greater the shift. This is another manifestation of the Doppler effect.

Around the time Einstein was developing his general theory, Vesto Slipher of Lowell Observatory in Arizona began measuring the radial velocities of spiral nebulae, including our neighbor Andromeda, which he found was approaching the Milky Way at a brisk 186 miles per second (roughly 300 kilometers per second).* Most of the dozens of spectra, however, were red-shifted: the galaxies were moving away from us. Furthermore, their velocities were huge, as high as 680 miles per second (1,100 kilometers per second), implying that the nebulae must reside well beyond the reach of the Milky Way's gravitational influence. It appeared Kant was right.

In 1912, the year that Slipher discovered galactic redshift, Henrietta Leavitt at Harvard College Observatory showed that there is a unique relationship between the luminosity and pulsation rate of *Cepheids*, types of stars whose brightness changes periodically. This relationship can be used to predict their distances. A little more than a decade later, Edwin Hubble found some Cepheids in Andromeda and used Leavitt's discovery to do just that. His measurements confirmed that the Cepheids—and, therefore, Andromeda—lie far outside the Milky Way, which conclusively settled the debate as to the extragalactic nature of this and other "spiral nebulae."

Hubble built upon this work by combining Leavitt's and Slipher's discoveries. He used Cepheids to determine the distances to more galaxies and combined these measurements with Slipher's redshift data. Bringing it all together, he found that the farther away a galaxy is, the more quickly it speeds away. This groundbreaking discovery was the first observational confirmation of an expanding universe. It marked a clear turning point in the history of astronomy and physics. To his credit, Einstein quickly abandoned his conviction that the universe was static, reflecting, "The redshift of the distant nebulae has smashed my old construction like a hammer blow."

When we say that the universe is expanding, we don't mean that galaxies are traveling *through* space, but that space itself is expanding. Moreover, if we lived in any galaxy other than the Milky Way, it would still appear as if all others were flying away from us. *There is no center to the universe.* Einstein understood these implications, and in addition to abandoning his idea of a static cosmology, he dispensed with his cosmological constant. Ever the

---

* Traveling at 186 miles per second, you would cross the state of Mississippi (width = 180 miles) in less than one second!

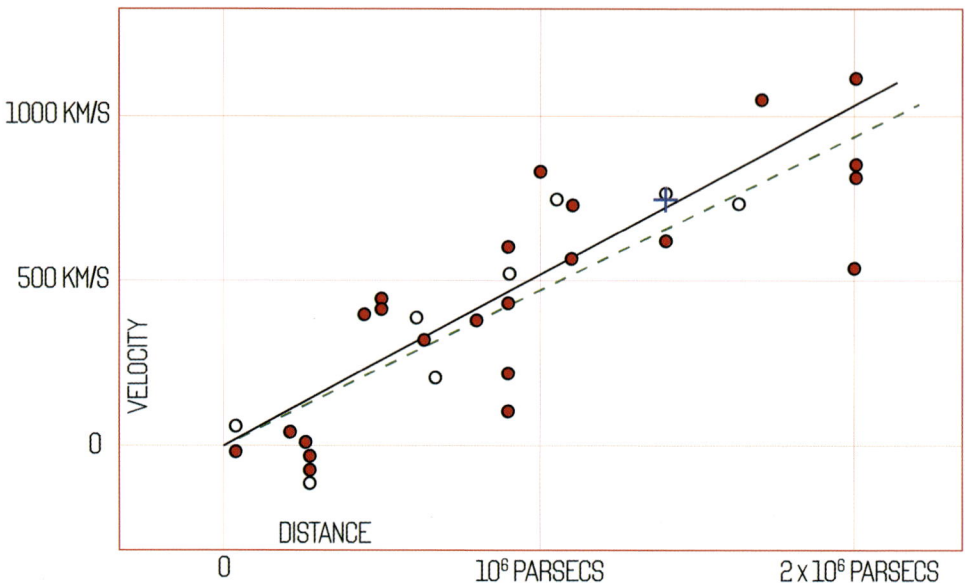

AN EXPANDING UNIVERSE | A color reproduction of Hubble's plot from his 1929 paper, which provides evidence that the universe is expanding. Each point represents the speed and distance he measured for a distant galaxy. A speed of 500 kilometers per second equals 1.1 million miles per hour. One parsec is equivalent to 3.26 light-years.

pragmatist, he no longer found it necessary in his new models of an expanding universe. Since general relativity could accommodate an expanding universe without the cosmological constant, he considered it redundant. According to physicist George Gamow, Einstein called it "my biggest blunder."

Nonetheless, decades later, when it was discovered that the universe is not only expanding but also accelerating, the cosmological constant made a sensational return to the stage. Today many scientists believe that Einstein's "biggest blunder" represents *dark energy*, an unknown force with a huge influence on the evolution of the universe.

Creative Faith

Experimenting, trying out new ideas, failing, learning from mistakes, trying again—trial and error—is inherent to the process of science. It takes faith to try new, uncertain pathways, especially when doing so goes against the established majority opinion. To make progress, scientists must have faith that the scientific process will eventually resolve errors and lead to truth. They must have some degree of confidence in the body of knowledge that forms the foundation, the starting point of their pursuits. They trust in the power of reason.

On a more personal level, faith plays a central role in the life of the individual scientist. To begin with, she must have faith in her vision. According to the psychologist Erich Fromm, important discoveries are rarely made by conducting arbitrary experiments and collecting data at random, without a vision in place. Distinguishing between *rational* and irrational faith, belief in something just because authority says so, Fromm writes, "At every step from the conception of a rational vision to the formulation of a theory, *faith* is necessary: faith in the vision as a rationally valid aim to pursue, faith in the hypothesis as a likely and plausible proposition, and faith in the final theory, at least until general consensus about its validity has been reached. This faith is rooted in one's own experience, in the confidence of one's power of thought, observation, and judgment."

One thing that sets the innovative scientist apart from the average one is the willingness to trust their experience, which includes their *intuition*. Trusting oneself in this way may involve confronting fears—of wasting time following an unproductive path, of failure, of embarrassment—and embracing the power of faith.

Faith is necessary because it is the catalyst for growth. Faith is a conviction about what is possible, and it requires imagination. As Voltaire said, "Faith consists of believing when it is beyond the power of reason to believe." And yet faith is not necessarily irrational. Faith takes humility because our reason is limited. An infant attempting to walk for the first time has no prior evidence to indicate she can do it. But without taking the risk to literally step out into the unknown, she will never grow into a fully developed human being. Faith leads to growth. Evolution without creative faith is impossible.

By *creative faith*, I mean a belief that has the potential to lead to growth and is backed up by action, a belief that isn't driven by fear (the opposite of faith). Fear keeps us stagnant when we give in to it. Many scientists and artists admit to feeling afraid when they set their mind to try something new or present their labor of love to an audience. Creation is a vulnerable act. You must overcome self-doubt to tread a new path. You open yourself up to criticism, or worse—indifference. You risk falling short of your own standards; you must be brave enough to confront your weaknesses and learn from your failures. Perhaps this is why French painter Henri Matisse said, "Creativity takes courage."

Like the history of science, the history of art is filled with examples of creative faith. Meta Vaux Warrick Fuller, Clementine Hunter, and Jean-Michel Basquiat were visual artists fully convicted of the power of art. Though they represented very different aspects of the black American experience, they shared a confidence in their own unique vision and tapped into this creative faith to overcome the doubts and social pressures that prevent most from realizing their potential. All were risk takers who in some way broke from convention and invented new artistic languages.

Sculptor, poet, and painter Meta Warrick Fuller was a giant of the Harlem Renaissance. Born in Philadelphia to a middle-class family, she was given educational opportunities that most other blacks didn't have access to. In 1899, the young artist went to study in Paris. While she didn't escape racism in Europe, she was able to blossom as an artist there. She was mentored by expatriate painter Henry Ossawa Tanner and also met Auguste Rodin, who took her on as a protégé. The great sculptor encouraged her to cultivate her gift of depicting human suffering with an uncanny visceral realism. When she returned to Philadelphia, the acclaim she had started to receive in Europe didn't lessen the particularly intense brand of American racism she confronted there once again. Despite the hostility, she blazed the trail on how Afrocentric themes are portrayed through sculpture. She drew inspiration from African history, as in the sculpture *Ethiopia*; from poetry, as in *Man Eating His Heart*, inspired by a line in Stephen Crane's poem "In the Desert"; and from current events, as in the sculpture of Mary Turner, a pregnant woman who was lynched in 1918.

Clementine Hunter was born about 10 years after Fuller, in 1886 or 1887, on a former plantation in north-central Louisiana. Hunter belonged

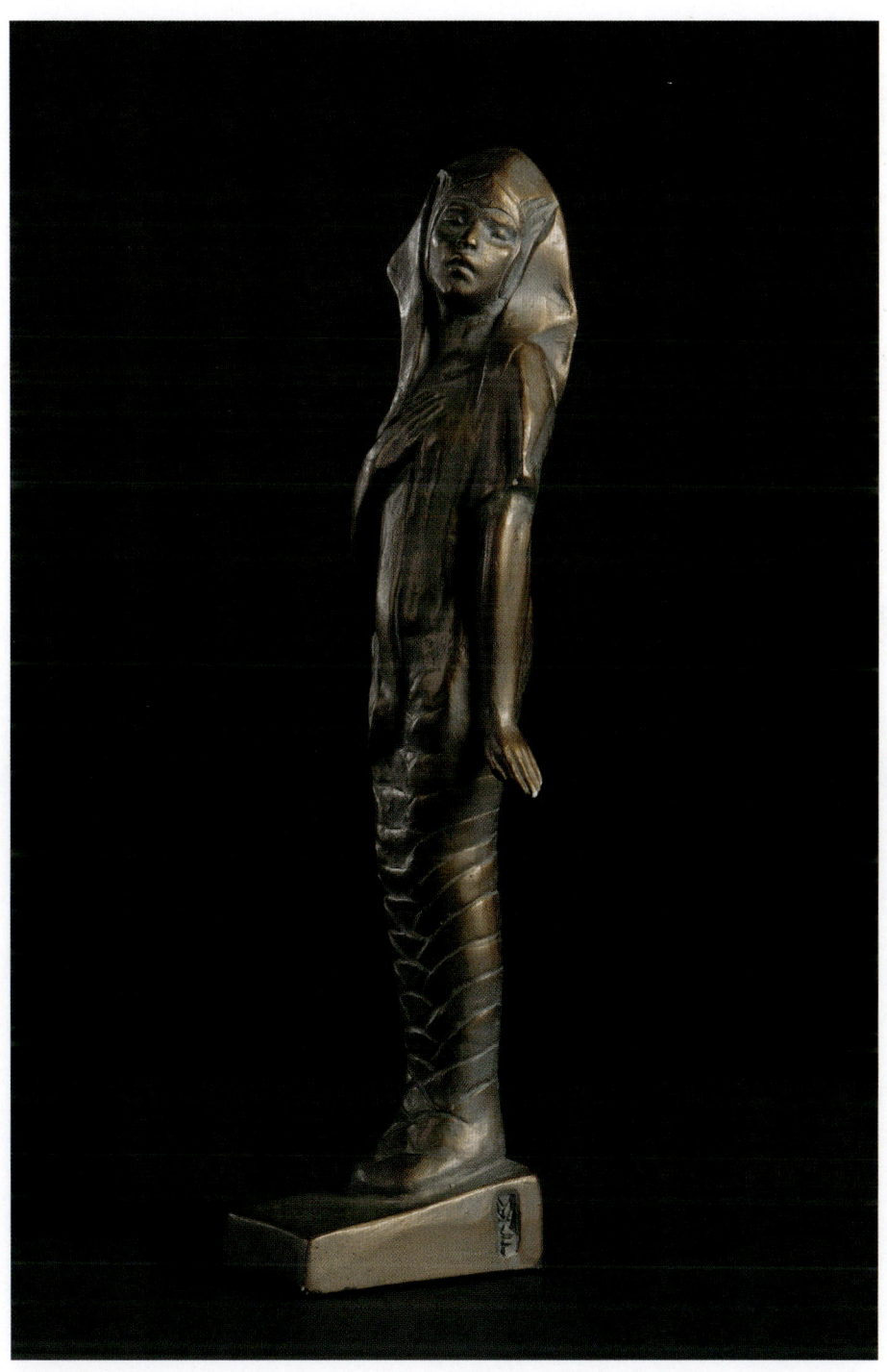

ETHIOPIA | Meta Vaux Warrick Fuller, ca. 1921, paint on plaster, 13 × 3½ × 3⅞ in (33 × 8.9 × 9.8 cm).

to a family of sharecroppers who raised cotton, sugarcane, corn, pecans, and other cash crops for wealthy planters. Many of her paintings depict scenes she remembered from the fields—rows of women with their arms outstretched toward the cotton plants, with long, heavy white sacks strapped around the chest. The women and the cotton plants fill the frame of the canvas, almost overflow it, with perhaps just a narrow strip of sky running across the top; the tight sense of space emphasizes the intimate relationship between the women, the land, and their work. Hunter's palette, always bright and colorful, might seem to belie the reality of this backbreaking work. Unlike Fuller, Hunter was self-taught. She didn't begin painting until her late forties, after her work shifted from the fields to the "big house." The white woman whose husband owned Melrose Plantation, where Hunter lived most of her life, was a patron of the arts, and it was here that Hunter was exposed to the "art world." She took to painting and began working almost incessantly, creating thousands of pieces over the next five decades of her life.

It's clear from Hunter's work that she was motivated by her belief in God, in an unseen spiritual world. She believed that angels always surround us, and her skies are often populated with crowds of them, winged white and red creatures, hair blowing in the wind as they fly across Earth. But her faith was far from irrational. She spoke to her confidence in the power of her thought (to paraphrase Fromm) when she said, "Painting is a lot harder than picking cotton. Cotton's right there for you to pull off the stalk, but to paint, you got to sweat your mind."

In Fuller's and Hunter's art, there is often a striking dissonance—a palpable tension between the subject matter and the beauty of form or vibrance of color of the art itself. The work of Jean-Michel Basquiat shares this quality of dissonance, a hallmark of much black art—visual art, music, dance, and literature—because it's such an integral part of the black experience, diverse as it is. Of the three artists discussed here, the Brooklyn-born Basquiat is the most famous and widely studied. Born in 1960 to parents from Haiti and Puerto Rico, both countries that have been traumatized by US intervention, the influential neo-expressionist artist is celebrated for revolutionizing contemporary art.

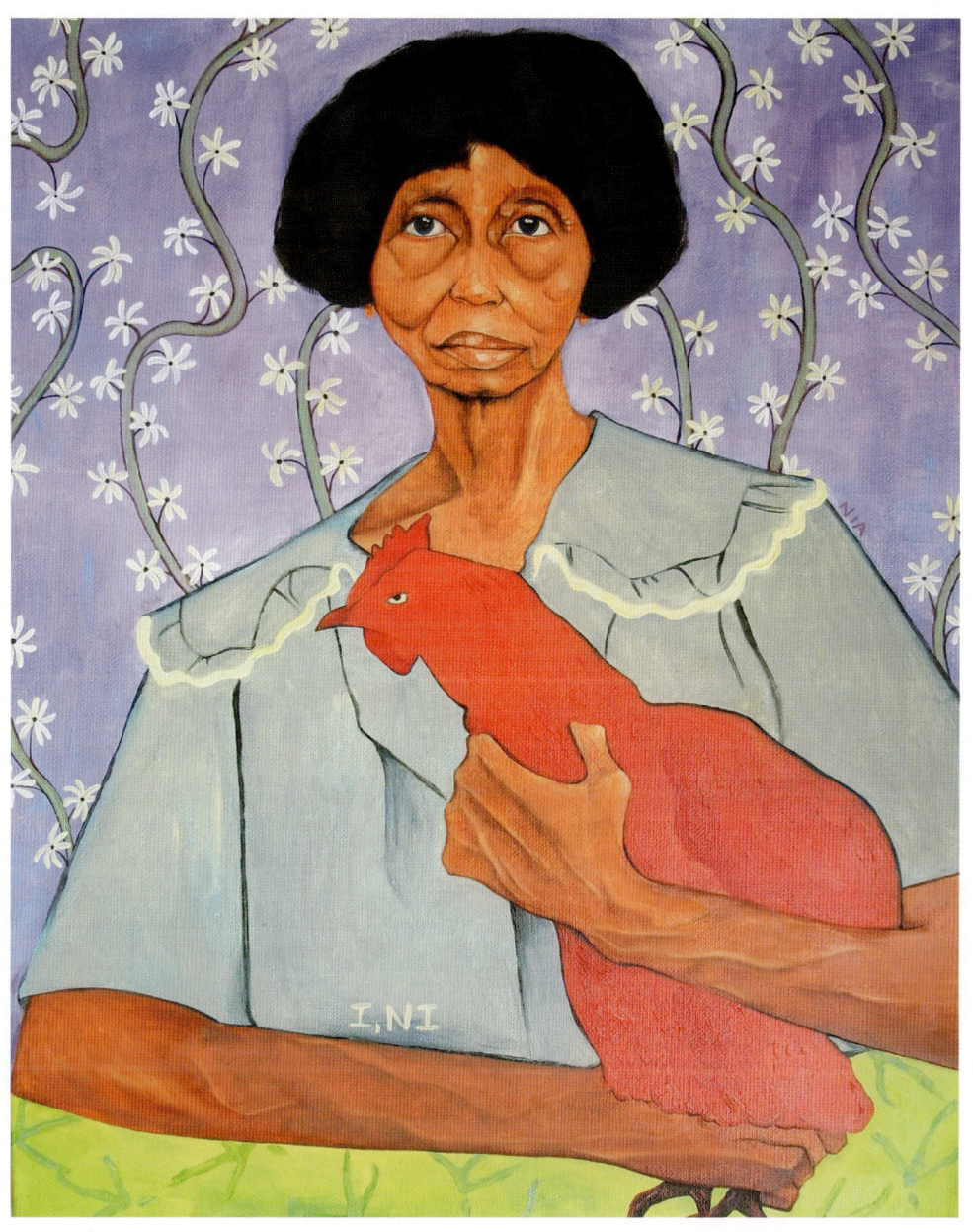

CLEMENTINE HUNTER | Nia Imara, 2014, oil on canvas, 16 × 20 in (40.6 × 50.8 cm).

I bring up Basquiat because he captures the essence of creative faith. His vast knowledge shines through in the dizzying range of references in his paintings—African art, European art, ancient history, black American history, current events, Caribbean culture, American pop culture, human anatomy, literature, hip-hop, jazz—all held together by his mastery of color, composition, symbolism, and storytelling. Over his brief career he developed a sophisticated analysis of racism and capitalism, even if this seemed to contradict his desire to be rich, famous, and accepted within the white art establishment. He used his virtuosic talent to honor the lives of black American heroes (in the painting *Horn Players,* for example), challenge how they are perceived (*Hollywood Africans*), and expose the contradictions of black life in a racist society (*Irony of a Negro Policeman*).

During Basquiat's life and even today, commentaries about his work often try to separate his technical virtuosity from the brilliant social message of his art. Sometimes this is done by minimizing or omitting any reference to racism. This isn't uncommon for prominent black artists who are subject to the scrutiny of a non-black audience that doesn't always understand the weight of the history, culture, and lived experience that informs their work. In all great art, technique and the deeper message reinforce each other. Without the sincerity and depth of Basquiat's social message, the work might be flashy and exciting, technically speaking, but empty; without the technical skill, the social message might be reduced to propaganda—or simply bad art. In any case, in both technique and social message, Basquiat pushed boundaries, stepped over them, demolished them completely in service of his vision. Such boldness and confidence in the worthiness of one's vision is impossible without faith.

Like the cosmos itself, the human universe of art is dynamic, responsive to invisible forces. Fuller, Hunter, and Basquiat may be counted among the great artists who, by reflecting humanity to itself in brand-new ways, went beyond responding to these forces to influencing them. Rather than be passive members in a social spacetime that rippled about them, they became active agents whose faith in action catalyzed its evolution.

"What the Senses Do Not Tell"

Fifty years after Hubble's discovery, scientists were still trying to understand why the universe is expanding. Their efforts to solve this mystery only deepened it.

In the 1990s, two independent research groups used Type Ia supernovae to trace the history of the universe's expansion. Type Ia supernovae, distinguished by their uniform luminosity, occur in binary star systems in which one of the stars is a white dwarf. As the white dwarf rips matter away from its nearby companion, it grows in mass. Since all white dwarfs attain the same mass before exploding, they all attain the same maximum luminosity, and astronomers can use this information to measure the distances to their host galaxies. The Supernova Cosmology Project, led by Saul Perlmutter, and the High-Z Supernova Search Team, cofounded by Brian Schmidt and Nicholas Suntzeff, compared the distances calculated in this way to distances they calculated using a different, independent technique. Both teams found a persistent discrepancy between the two sets of distance measurements. Their analyses implied that the supernovae were actually located farther away than anticipated. This led to a striking conclusion: rather than expanding at a constant rate, the universe's expansion is actually *accelerating*!

What driving force is responsible for pushing the universe apart at an ever-increasing rate? The most widely accepted explanation is *dark energy*, a sort of antigravity evenly distributed throughout space. The evidence for its existence is persuasive, yet we have no idea what it actually is. Some scientists theorize that it is a form of energy intrinsic to the fabric of spacetime—or perhaps a fifth fundamental force of nature.

Many scientists believe that dark energy is the cosmological constant, the fudge factor that in his theory of general relativity Einstein referred to by the Greek letter "lambda" (Λ). For Einstein, who believed the universe was static at the time he developed the cosmological constant, Λ represented a hypothetical source of repulsion that could balance gravity and prevent the universe from contracting or expanding. Modern cosmologists take Λ to represent vacuum energy, the intrinsic energy of "empty" space. The problem is that the energy of empty space is much, much smaller than what most theoretical predictions say Λ should be. In fact, the discrepancy is 120 orders of magnitude. This is like saying that the mass of an electron (the lightest subatomic particle) is actually far more massive than the entire known universe! This huge disparity has led some to call calculations of Λ "the worst prediction in the history of physics."

Whatever dark energy is, the acceleration of the universe did not begin at the Big Bang. For the first several billion years, gravity dominated and was powerful enough to curb expansion. It was not until about 9.8 billion years after the Big Bang that dark energy began to supersede gravity, and the rate of the universe's expansion started to increase over time.

Before the discovery of dark energy, scientists developed hypothetical models of the universe's evolution. In these models, the universe grows or diminishes in size over time, and its evolution depends on how much mass it contains as well as the rate of expansion. To get a sense of how much mass there is, astronomers measure the average density in large volumes of space. If the density is very low, gravity will not be strong enough to slow the stretching of space, and the universe can grow forever. A high density of matter would correspond to greater gravity, which could lead to deceleration and even cause the universe to collapse back on itself into a "Big Crunch," as physicist John Wheeler called it. But if the average density of the cosmos is equal to a certain critical value, expansion will gradually halt at some time in the infinite future.

The theory of general relativity tells us that matter produces the curvature of space. This means that the average density of the universe is tied to its overall geometry. Prior to Einstein, the idea that the universe could have a geometry or shape would have seemed absurd. But we now know that the global geometry of the universe is an essential property that determines its evolution.

There are at least three possible geometries of the universe: open, closed, or flat. In a closed universe, spacetime might resemble the surface of a sphere. In such a universe, two rays of light that started out parallel would eventually join.

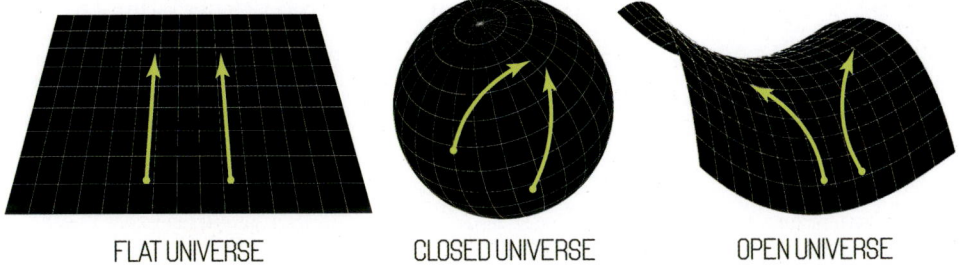

FLAT UNIVERSE   CLOSED UNIVERSE   OPEN UNIVERSE

GEOMETRIC MODELS OF THE UNIVERSE | In a flat universe, two beams of light that start out parallel will remain parallel. In a closed or open universe, light beams starting out parallel will eventually converge or diverge, respectively. (Keep in mind that the universe is four-dimensional, while the analogies shown here are two-dimensional.)

A closed geometry would result if the average density of the cosmos exceeded a certain amount known as the *critical density*, in which case gravity would eventually slow down expansion, cause it to halt, and finally cause everything to draw back in upon itself. In an open universe, spacetime could resemble the shape of a horse's saddle; two initially parallel light rays would deviate. Such a universe would be possible if the average density were less than the critical value, and cosmic expansion would coast indefinitely, neither decelerated by gravity or accelerated. In a flat universe, spacetime is like our flat, extended sheet described earlier. In this scenario, the average density of the universe just balances the critical density, so that it will go on expanding forever.

Agreement from independent experimental efforts leads most astronomers to expect that we live in a flat universe. In particular locations, such as around massive objects like stars and galaxies, space experiences curvature, but overall, the universe as a whole may be flat. Kind of like an expansive, flat desert that has some occasional shrubs and stones here and there. Theory predicts that the critical density of the universe today is about *six protons per cubic meter.** When astronomers tabulate all of the ordinary visible matter found in

---

* One cubic meter is equivalent to a box whose dimensions are 1 meter (1.1 yards) on each side.

COMPOSITION of the UNIVERSE

"ORDINARY"
MATTER
(5%)

STARS

DARK MATTER
(27%)

DARK ENERGY
(68%)

stars, galaxies, and intergalactic space, this amounts to only 5 percent of the predicted critical density. The contribution from dark matter adds another 27 percent. Where is the rest? The answer: dark energy. When we use $E = mc^2$ to calculate the mass equivalent of dark energy, we find that it contributes to the remaining 68 percent of the critical density. In other words, the vast majority of the stuff of the universe consists of an invisible form of matter and a cosmic antigravity that are complete mysteries to us!

The most compelling physical evidence that the universe is spatially flat comes from what is known as the cosmic microwave background (CMB), a very low-energy radiation that permeates the universe in all directions. The CMB was discovered serendipitously in 1964, when astronomers Robert Wilson and Arno Penzias realized that their Bell Labs radio telescope was picking up a faint buzzing sound—like radio static—coming from all parts of the sky. Penzias and Wilson did everything they could to identify and eliminate the source of noise, even removing some pigeons they found roosting in the antenna, but the hum persisted. The importance of their discovery became apparent when they began talking to cosmologists, whose models of the early universe predicted an "echo" of the Big Bang permeating space.

To understand the origins of the CMB and its significance, let's go back to the very beginning. The early universe was a hot, dense plasma containing mostly light. It was so hot and energetic that for the first 10^{-43} seconds, the four fundamental forces of nature were unified. The force of gravity, electromagnetism (light), the strong force, and weak force were one.* With such high energies, conditions were favorable for light to spontaneously convert into particles and antiparticles, which would almost instantaneously self-annihilate and produce radiation. At 0.01 seconds, the temperature of the universe was about 100 billion Kelvin, still too hot for the universe to be anything other than a gumbo of radiation, protons, neutrons, and electrons.

---

* The strong and weak forces work on the subatomic level. The strong force holds together the nuclei of atoms, while the weak force is responsible for the radioactive decay of particles.

By the time the universe was three minutes old, the temperature fell to one billion Kelvin; at four minutes, the nucleus of a helium atom could form.

As the universe expanded, it continued to cool. Yet for the first several thousand years, it remained so hot that atoms were ionized, so that there were lots of free electrons flying around. This meant that photons could not travel very far before bouncing off a free electron. In other words, light could not stream freely through the universe. The universe was opaque. By its 380,000-year birthday, the universe had cooled down to 3,000 Kelvin—cool as the surface of a low-mass star—which meant that electrons and nuclei could combine for the first time to form neutral hydrogen and helium. At this point, light was able to freely travel through the universe. For the first time, the lights were turned on.

When we look outward into space, we can see this first "afterglow" of the early universe. That's what the cosmic microwave background radiation is. Due to the stretching of space since its birth more than 13 billion years ago, the CMB has been redshifted to low-energy radiation having an average temperature of 2.7 Kelvin, the coldest naturally occurring phenomenon in the cosmos. Scientists have been able to capture images of the CMB with telescopes tuned to its frequency. These images are iconic for astronomers; they are the earliest photos we have of the infant universe. The most recent dappled image shows that the temperature is remarkably uniform, with tiny fluctuations on the order of one part in 100,000. Slightly warmer spots, indicated by the color red, represent regions of space in the early universe that had slightly lower than average densities. Slightly colder spots, shown in blue, correspond to higher densities that evolved into the galaxy clusters, galaxies, and stars we observe today. Theory predicts that the sizes of the hot and cold spots in the CMB are determined by the geometry of the universe and that they are consistent with the idea that the cosmos was flat from very early in its history.

Like the cosmic Goldilocks seeking the perfect bowl of porridge, if the universe had had slightly more matter, its expansion would have eventually been reversed and it would have collapsed. If the universe had been just a little less dense, it would have inflated too quickly for structure to develop.

Luckily for us, we live in a "just right" universe. Had the cosmos not been so finely balanced from the very beginning, we would not be here.

We've come this far by faith.
— Albert A. Goodson —

That's what science tells us anyway. The deeper part of me believes that the soul is eternal, that as unlikely as our existence seems, we were meant to be here. I have faith that whether the universe has four dimensions—or 11, or 26, like the string theorists say—that love exists beyond space and time. Physics tells us there are four forces of nature, but I know there's a force greater than all of these. As the nature of gravity is to bring things together, it is our nature to love and receive love. I have faith in the creative power of love—the matrix of the cosmos.

How, then, does all the tragedy and suffering of human life fit into this matrix? Where is the love for Pecola Breedlove? She's a fictional character but she's real; her story is true and has happened millions of times and is happening still. What do we make of a universe that permits the conditions for the existence of Pecola Breedloves? The ending of *The Bluest Eye* frightens me. I hope it is not too late. I have faith that though Toni Morrison ended the child's story, somehow it isn't really finished. It is a mystery more awesome than dark energy. I understand little of it now, but, like the woman at the well, I expect that one day all will be revealed. But even if it isn't, I have faith that life is good. I have faith that the love we create creates us. I look up and live in excited expectation. We are alive! We can grow; we can love! It is good!

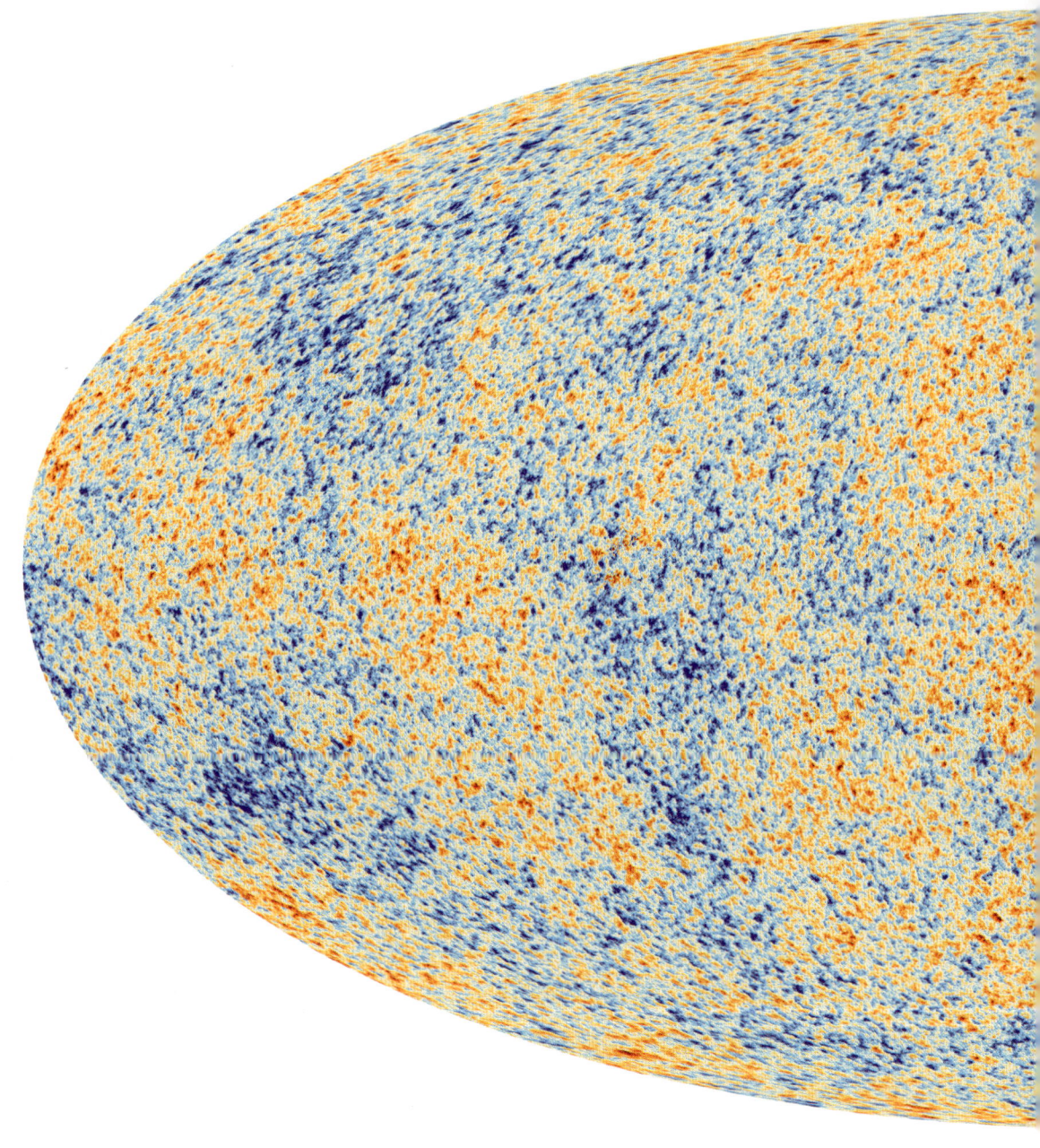

OLDEST LIGHT IN THE UNIVERSE | All-sky map of the cosmic microwave background.

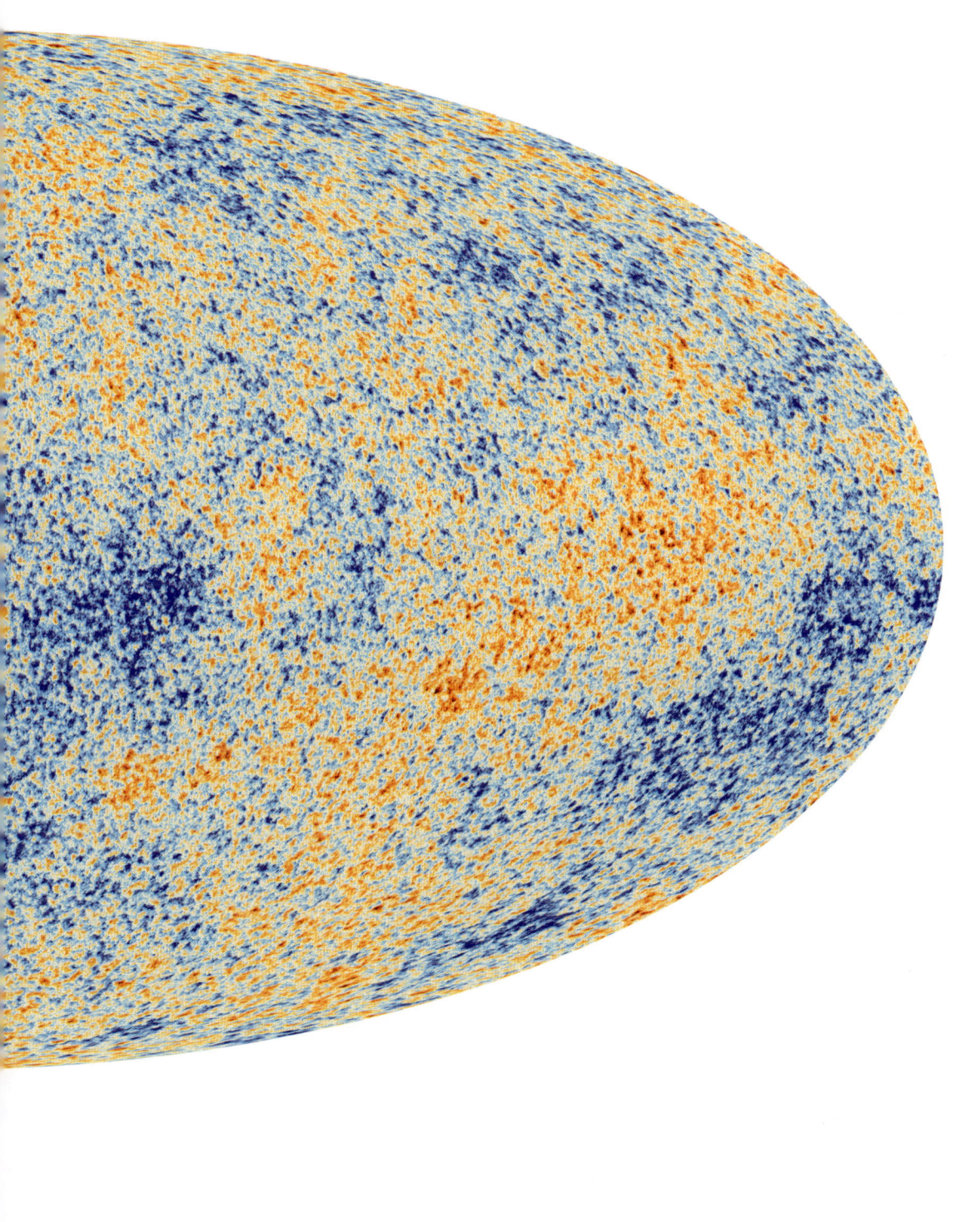

Acknowledgments

Lauren Bieker—for your faithfulness, encouragement, consistency, and kindheartedness—thank you.

Ian Weaver, I'm grateful for your indispensable assistance securing permissions for the images.

Anna Pancoast, thank you for your help conceiving the original illustrations for this book, for the conversations that sparked new and better ideas. Thank you for your friendship.

Image Credits

Chapter One

Wonder of the World (page 5)—Image courtesy of the Earth Science and Remote Sensing Unit, NASA Johnson Space Center. NASA Photo ID ISS032-E-9123; eol.jsc.nasa.gov.

Cross Section of Giza (page 7)—Image by Nia Imara.

Rhind Papyrus (page 8)—Image licensed from the British Museum.

Ancient Art (page 11)—Image courtesy of Professor Christopher Henshilwood.

Nabta Playa (page 13)—Image by Raymbetz, licensed via Creative Commons Attribution-Share Alike 3.0 (https://creativecommons.org/licenses/by-sa/3.0/), from https://commons.wikimedia.org/wiki/File:Calendar_aswan.JPG.

Goddesses of Knowledge (page 16)—*Left:* Seshat image from "Temple of Luxor" photos by Martin Doudoroff, licensed via Creative Commons Attribution 2.0 Generic (https://creativecommons.org/licenses/by/2.0/), from https://www.flickr.com/photos/doudoroff/52785650005/. *Top right:* Nisaba image by Osama Shukir Muhammed Amin FRCP(Glasg), licensed via Creative Commons Attribution-ShareAlike 4.0 International (https://creativecommons.org/licenses/by-sa/4.0/deed.en), from https://commons.wikimedia.org/wiki/File:Sumerian_goddess_Nisaba,_the_name_of_Entemena_is_inscribed,_c._2430_BC,_from_Southern_Mesopotamia,_Iraq.jpg.

The Banjo Lesson (page 25)—Image from Hampton University Museum / the Artchive.

George Washington Carver (page 27)—Image by Arthur Rothstein, from the Collection of the Smithsonian National Museum of African American History and Culture.

Chapter Two

Untitled (page 32)—*Untitled;* Gavin Jantjes, born 1948, South Africa; Acrylic on canvas; H x W: 200 x 300 cm (78¾ x 118⅛ in.); 96-23-1; © 1989–90 Gavin Jantjes; Purchased with funds provided by the Smithsonian Collections Acquisition Program; Photograph by Franko Khoury, National Museum of African Art, Smithsonian Institution. Also © 2024 Artists Rights Society (ARS), New York / DACS, London.

The Moon, Mars, Saturn, and Jupiter Together (page 35)—Image by Mihail Minkov.

The Dresden Codex (page 36)—Image courtesy of SLUB Dresden / Mscr. Dresd.R.310.

An Unexpected Guest (page 42)—Image via the Astrophysics Data System.

Crab Nebula (page 43)—Most detailed image of the Crab Nebula (image ID heic0515a) by NASA, ESA and Allison Loll/Jeff Hester (Arizona State University). Acknowledgment: Davide De Martin (ESA/Hubble). Licensed via Creative Commons Attribution 4.0 International (https://creativecommons.org/licenses/by/4.0/), from https://esahubble.org/images/heic0515a/.

Sagittarius A* (page 45)—Image via Chandra X-ray: NASA/CXC/SAO; IR: NASA/HST/STScI.

The Andromeda Galaxy (page 48)—Image courtesy of Robert Gendler.

The Virgo Cluster (page 49)—Image courtesy of Dr. Fernando Peña Campos.

Hubble Ultra Deep Field (page 50)—Hubble Ultra Deep Field (image ID heic0611b) by NASA, ESA, and S. Beckwith (STScI) and the HUDF Team, licensed via Creative Commons Attribution 4.0 International (https://creativecommons.org/licenses/by/4.0/), from https://esahubble .org/images/heic0611b/.

Chapter Three

Ibn al-Haytham's "dark room" experiment (page 55)—Image by Nia Imara.

Book of Optics (page 56)—Image ID: 2B02PH6 licensed from CPA Media Pte Ltd / Alamy.

Light waves (page 58)—Image by Nia Imara.

The Electromagnetic Spectrum (page 60)—Image by Nia Imara.

Newton's Prism Experiment (page 63)—Image by Nia Imara.

How a Rainbow Forms (page 65)—Image by Nia Imara.

Newton's Color Wheel (page 68)—Image via Wikimedia Commons.

Goethe's Color Wheel (page 69)—Image via Wikimedia Commons.

Mother and Child (page 70)—© 2024 Estate of Pablo Picasso / Artists Rights Society (ARS), New York.

Net Casting (page 71)—© Jonathan Green.

Falling Star (page 75)—© 2024 Romare Bearden Foundation / Licensed by VAGA at Artists Rights Society (ARS), New York.

Circe (page 76)—the Painter of the Boston Polyphemos. Drinking cup (kylix) depicting scenes from the *Odyssey*. Greek, Archaic Period,

550–525 BC. Place of Manufacture: Greece, Attica, Athens. Ceramic, Black Figure. Height: 13.2 cm (5³⁄₁₆ in.); diameter: 21.7 cm (8⁹⁄₁₆ in.). Museum of Fine Arts, Boston, Henry Lillie Pierce Fund, 99.518.

Untitled ("Black Circe") (page 77)—© 2024 Romare Bearden Foundation / Licensed by VAGA at Artists Rights Society (ARS), New York.

Pilate (page 78)—© 2024 Romare Bearden Foundation / Licensed by VAGA at Artists Rights Society (ARS), New York.

Chapter Four

Shout (page 83)—Image by Nia Imara.

Stela of Aafenmut (page 84)—Image from the Met Museum via Open Access.

Gold headdress ornament (page 84)—Image from the Met Museum via Open Access.

Sun and Plum Branches (page 85)—Image from the Met Museum via Open Access.

Mandala of the Sun God Surya (page 85)—Image from the Met Museum via Open Access.

Coronation Stone of Motecuhzoma II (Stone of the Five Suns) (page 85)— Aztec (Mexica), 1503. The Art Institute of Chicago.

The Active Sun (page 87)—Image courtesy of Mark Johnston (@azastroguy).

The Eclipse (page 90)—© 2024 Estate of Alma Thomas (Courtesy of the Hart Family) / Artists Rights Society (ARS), New York; Smithsonian American Art Museum, Washington, DC / Art Resource, New York.

Nwantantay (page 92)—Image ID:P2R5KN, licensed from Hemis / Alamy.

A Bwa Ceremony (page 93)—Image ID:2CC4NTM, licensed from Elena Bobrova / Alamy.

Cross Section of the Sun (page 97)—Image by Nia Imara.

Total Eclipse (page 98)—Image courtesy of Phil Hart.

Sounds of the Sun (page 100)—Illustration of Sun's vibrational modes (image ID noao0002a) by NSO/AURA/NSF, licensed via Creative Commons Attribution 4.0 International (https://creativecommons.org /licenses/by/4.0/), from https://noirlab.edu/public/images/noao0002a/.

Chapter Five

Exoplanet Zoo (page 110)—Image by Martin Vargic, www.halcyonmaps .com.

Beta Pictoris b (page 112)—Beta Pictoris as seen in infrared light (image ID eso0842a) by ESO/A.-M. Lagrange et al., licensed via Creative Commons Attribution 4.0 International (https://creativecommons .org/licenses/by/4.0/), from https://www.eso.org/public/images/eso 0842a/.

Transits (page 113)—Image by Nia Imara.

Kepler Space Telescope (page 114)—Image via NASA/JPL.

TRAPPIST-1 (page 116)—Illustration: NASA/JPL-Caltech.

Life on Proxima b (page 118)—Artist's impression of the planet orbiting Proxima Centauri (image ID eso1629a) by ESO/M. Kornmesser, licensed via Creative Commons Attribution 4.0 International (https:// creativecommons.org/licenses/by/4.0/), from https://www.eso.org /public/images/eso1629a/.

Drake Equation (page 124)—Image by Nia Imara.

Jill Tarter (page 125)—Photograph by Louis Psihoyos, courtesy of Jill Tarter.

Improvisation No. 30 (Cannons) (page 127)—Wassily Kandinsky, 1913. The Art Institute of Chicago.

Chapter Six

Gravitational Waves (page 136)—Image via S. Ossokine, A. Buonanno (Max Planck Institute for Gravitational Physics), Simulating eXtreme Spacetimes project, D. Steinhauser (Airborne Hydro Mapping GmbH).

First binary black hole merger detected by LIGO (page 137)—Image courtesy of Aurore Simonet.

A Stellar Mass Black Hole (page 141)—Image via NASA/CXC/M.Weiss.

Center of the Milky Way (page 142)—Image via Ian Heywood (Oxford U.), SARAO; Color Processing: Juan Carlos Munoz-Mateos (ESO).

Sagittarius A* (page 147)—First image of our black hole (image ID eso2208-eht-mwa) by EHT Collaboration, licensed via Creative Commons Attribution 4.0 International (https://creativecommons.org/licenses/by/4.0/), from https://www.eso.org/public/images/eso2208-eht-mwa/.

Harlem Woman (page 149)—© 2024 Mora-Catlett Family / Licensed by VAGA at Artists Rights Society (ARS), New York.

In Sojourner Truth . . . (page 151)—*In Sojourner Truth I fought for the rights of women as well as Negroes.* 1947, printed 1989. From the Portfolio: *The Negro Woman*, 1946-47 (re-titled *The Black Woman*, 1989). Edition: 14/20 | Second print run. Printed by Robert Blackburn Printmaking Workshop. Linoleum cut, Sheet (Irregular): 10¼ × 7⅜ in. (26 × 18.7 cm) Image: 8⅞ × 5¹⁵⁄₁₆ in. (22.5 × 15.1 cm). Purchase, with funds from the Print Committee, Inv No. 95.195. Digital image © Whitney Museum of American Art / Licensed by Scala / Art Resource, New York and © 2024 Mora-Catlett Family / Licensed by VAGA at Artists Rights Society (ARS), New York.

In Harriet Tubman . . . (page 154)—*In Harriet Tubman I helped hundreds to freedom.* 1946, printed 1989. From the Portfolio: *The Negro Woman*, 1946–47 (re-titled *The Black Woman*, 1989). Edition: 13/20 | Second print run. Printed by Robert Blackburn Printmaking Workshop. Linoleum cut, Sheet (Irregular): 10¼ × 7¾ in. (26 × 19.7 cm) Image (Irregular): 9⅛ × 7¹⁄₁₆ in. (23.2 × 17.9 cm). Purchase, with funds from the

Print Committee, Inv No. 95.194. Digital image © Whitney Museum of American Art / Licensed by Scala / Art Resource, New York and © 2024 Mora-Catlett Family / Licensed by VAGA at Artists Rights Society (ARS), New York.

The Waterfall Where Yoshitsune . . . (page 155)—Image from the Met Museum via Open Access.

Augusta Savage (page 156)—Andrew Herman. Augusta Savage with her sculpture *Realization*, circa 1938. Federal Art Project, Photographic Division collection, circa 1920–1965. Archives of American Art, Smithsonian Institution.

The Death of Cleopatra (page 157)—Image from the Smithsonian American Art Museum. Gift of the Historical Society of Forest Park, Illinois.

Chapter Seven

My imagining of . . . (page 162)—Image by Nia Imara, collage of Vincent van Gogh, *Self-Portait* (1887), image from the Joseph Winterbotham Collection at the Art Institute of Chicago; Vincent van Gogh, *Olive Trees* (1889), image from the Met Museum via Open Access; and Paul Gauguin, *Portrait de l'artiste* (1893/1894), © RMN-Grand Palais (Musée d'Orsay) / Franck Raux.

Allegory of Time Governed by Prudence (page 166)—Image from the National Gallery (UK), presented by Betty and David Koetser, 1966.

Le Déjeuner sur l'herbe: Les Trois Femmes Noir (page 167)—Image licensed from Mickalene Thomas, LLC.

Le Déjeuner sur l'herbe (page 168)—Image from the Musée d'Orsay.

The Judgment of Paris (page 169)—Image from the Met Museum via Open Access.

Souvenir I (page 170)—© Kerry James Marshall, 1997. Courtesy of the artist and Jack Shainman Gallery, New York.

The oldest discovered sundial (page 174)—Image from University of Basel.

Chapter Eight

Aspiration (page 179)—© 2024 Heirs of Aaron Douglas / Licensed by VAGA at Artists Rights Society (ARS), New York.

Let My People Go (page 183)—© 2024 Heirs of Aaron Douglas / Licensed by VAGA at Artists Rights Society (ARS), New York and © The Metropolitan Museum of Art. Image source: Art Resource, New York.

Planetary Music (page 187)—Image from the University of Oklahoma.

Saturn (page 190)—Image via NASA/JPL-Caltech/Space Science Institute.

Venus Crossing the Sun (page 192)—Photograph by David Cortner.

Chapter Nine

Plantation (page 199)—© Estate of Elizabeth Talford Scott at Goya Contemporary/TALP.

Overview of the Divine Comedy (page 201)—Image from Cornell University, PJ Mode Collection of Persuasive Cartography.

Galileo's observations of the planets (page 207)—Image from Istituto Di Linguistica Computazionale.

Galileo's drawings of the Moon (page 208)—Image via Wikimedia Commons.

A View of the Milky Way (page 209)—Image via JSTOR Open Collections.

Gaia map of the Milky Way (page 212)—© ESA/Gaia/DPAC; CC BY-SA 3.0 IGO. Acknowledgement: A. Moitinho.

The Andromeda Galaxy (page 215)—From Vera C. Rubin and W. Kent Ford, Jr., "Rotation of the Andromeda Nebula from a Spectroscopic Survey of Emission Regions," *Astrophysical Journal* 159 (February 1970): 379. © AAS. Reproduced with permission.

Rotation curve of Andromeda (page 216)—Image by Nia Imara, adapted from Rubin and Ford, "Rotation of the Andromeda Nebula from a Spectroscopic Survey of Emission Regions."

SMACS 0723 (page 218)—Image via NASA, ESA, CSA, STScI.

Simulating dark matter (page 219)—Image via Millennium Simulation (Springel et al., 2005).

Pictorial Quilt (page 223)—Pictorial quilt. American (Athens, Georgia). 1895–98. Object Place: Athens, Georgia, United States. By: Harriet Powers, American, 1837–1910. Cotton plain weave, pieced, appliqued, embroidered, and quilted. 175 x 266.7 cm (68⅞ x 105 in.). Museum of Fine Arts, Boston. Bequest of Maxim Karolik.

Chapter Ten

The Assassination of Pecola Breedlove (The Bluest Eye) (page 229)—© 2024 Emory Douglas / Licensed by AFNLYAW.com.

Bending of Starlight Near the Sun (page 233)—Image by Nia Imara.

An Expanding Universe (page 236)—Image by Nia Imara.

Ethiopia (page 239)—Collections of the Smithsonian National Museum of African American History and Culture, Gift of the Fuller Family, © Meta Vaux Warrick Fuller.

Clementine Hunter (page 241)—Image by Nia Imara.

Geometric Models of the Universe (page 245)—Image by Nia Imara.

Composition of the Universe (page 246)—Image by Nia Imara.

Oldest Light in the Universe (page 250)—© ESA and the Planck Collaboration.

Nia Imara (page 273)—Photograph by Arman Turner, courtesy of Nia Imara.

Index

A

Abena, 128–129
Aboriginal people, Australia, 12, 40
absorbed light, 63, 64. *see also* light
African Commune of Bad Relevant
    Artists, 225
Ailey, Alvin, 86
Albers, Josef, 73
Al-Biruni, 174, 175
Aldebaran, 14
alienation, sense of, 52
aliens, 121–126
Allegory of Time Governed by Prudence
    (Titian), 159, 165, 166
Allen Telescope Array, 121, 123
Almagest (Ptolemy), 200
Alpha Centauri, 38–39
Al-Sufi, 211
The Analects (Confucius), 184–185
ancient China, 42, 173. *see also* China
ancient Egypt, 16, 31, 39, 84, 172–
    174, 180–181. *see also* Egypt
ancient Greece, 15–17, 54–55, 173
ancient Rome, 40, 173
Andromeda galaxy, 47, 48, 214–217,
    234–235
Angelou, Maya, 81
antimatter, 194
Arecibo, 123
Aristarchus, 200, 202
Aristotle, 198
Armstrong, Louis, 89
art. *see also specific works and artists*
    abstract, 10, 22, 41, 91–93, 128,
        148–150
    as allegorical, 165
    defining, 21–26
    elements of visual art, 178–181

as essential to human experience,
    10
European ideas about, 14
faith in, 228
fields of, 17
harmony in, 178–184
intersection of science and, x–xii,
    3–28
making the invisible visible in,
    146, 148
as metaphor, 102–103
power of, 26
representational, 11–14
rhythms in, 89–91
science in, 21
separation of science and, 3–4
space in, 146, 148–158
spirituality and, 23, 27, 126–129,
    128, 240
subjective, 20
systems of, 26
as time machine, 176
Asante, 221
Aspiration (Douglas), 178–182, 184
*The Assassination of Pecola Breedlove
    (The Bluest Eye)* (Douglas), 229
astrobiology, 115
Astronomia Nova (New Astronomy)
    (Kepler), 186
astronomical clocks, 12–13, 30
astronomy. *see also individual topics*
    early ancestors' knowledge of,
        12–14
    and invention of the telescope,
        205–209
    storytelling based on (*see* stories)
astrophysics, 24, 41, 194
Athena, 17

Atum, 31
Augustine, Saint, 198
Azbuka (Tolstoy), 94
Aztecs, 85

B

Babylonians, 173, 174
Bach, J. S., 89
Bacon, Francis, 211
The Banjo Lesson (Tanner), 25
Banneker, Benjamin, 22, 181
Basquiat, Jean-Michel, 238, 240, 242
Bearden, Romare, 74–79, 152, 225
Bell Burnell, Jocelyn, 108
Benben, 7
Beta Pictoris b, 112
Big Bang, 51, 193–194, 244, 247
Big Dipper, 41
Bighorn medicine wheel, 13–14
biosignatures, 115
black holes, 44–46, 133–147
    binary black holes, 135–138
    definition of, 139
    discovery of, 135–139
    evolution of stars into, 139–144
    formation of, 133–134
    intermediate-mass, 139
    masses of, 138
    stellar-mass, 138–144
    supermassive, 138–139, 141,
        144–147
    ultramassive, 144
A Black Odyssey (Bearden), 76
"Blame It on the Sun" (Wonder), 89
Blombos Cave, 10–11
Blues (Catlett), 153
blues (music), 86
blueshift, 234

The Bluest Eye (Morrison), 227–228, 249
Boff, Leonardo, 82n
Book of Optics (Ibn al-Haytham), 56, 57
Bose, Satyendra Nath, 23
Brackens, Laverne, 224
Brahe, Tycho, 203–205
Brown, James, 89
brown dwarves, 121
Buller, A. H. Reginald, 164n
Butler, Bisa, 225
Butler, Octavia, 29
Bwa people, 14, 91–94. *see also* Nwantantay

C

Caetani, Michelangelo, 201
calendars, 13–14, 30, 171–172
Calima cultures, 84
Callisto, 206
Carver, George Washington, xi, 26–28, 130, 181
Cassiopeia, 203
Catlett, Elizabeth, 148–154, 156, 158
cave art, 10–12
Cepheids, 235
Cepheus, 7
Cézanne, Paul, 24, 72, 177
Chandrasekhar, Subrahmanyan, 140
China
    ancient, 42, 173
    astronomy, 42, 160
    calendar, 173
    context of colors, 73, 79
chromosphere, 97, 99
Circe, 74, 76–77, 79
Circe (Bearden), 74, 76–77
Clark, Mark, 170
Clementine Hunter (Imara), 241
clockwork universe, 18
closed universe, 245
Collins, Addie Mae, 170, 171
color(s), 57–80
    color theory, 67, 72–74
    color wheel, 68–69
    feelings and emotions engendered by, 74–80
    harmony, 180
    human perception of, 59–63
    and interactions of light with matter, 63–66
    Kandinsky on, 126
    as light, 62
    meaning of, 67–74
    Munsell's color tree, 72
    primary, 73

racist interpretations of black, 79–80
rainbows, 65
relationships between, 73
and rhythm, 89–91
symbolism, 67, 73, 79–80
A Color Notation (Munsell), 72
Coltrane, John, xi, 86
Coma Cluster, 41, 217
conceptions of the cosmos, 197–225
    and dark matter, 215–220
    geocentric, 198, 200, 202
    heliocentric, 200, 202–205
    and humility, 197–198, 220–221
    and invention of telescopes, 205–209
    and planetary movements, 200, 202, 204
"Concerning the Spiritual in Art" (Kandinsky), 126, 128
Confucius, 184–185
constellations, 12, 33, 120. *see also* individual constellations
Contact (Sagan), 123
Copernicus, Nicolaus, 202, 203, 225
corona, 97, 99
Corona Borealis, 41
Coronation Stone of Motecuhzoma II (Stone of the Five Suns), 85
cosmic microwave background (CMB), 247–248, 250–251
cosmological constant, 232, 235, 243
cosmos. *see also* universe
    average density of, 244, 245
    black holes in, 144
    clockwork, 18
    communication with, 7
    conceptions of (*see* conceptions of the cosmos)
    and concept of spacetime, 134
    cosmic microwave background radiation in, 248
    and cosmological constant, 233
    dark matter in, 46, 106
    elements for life in, 44
    exploring workings of, 24
    harmony of, 185
    history of, 39
    humanity's experience of, 191, 193
    humans' relationship to, 13, 31, 148, 184, 198, 205, 214
    humans' understanding of, 40, 51–52
    life in the, 107, 115, 121–122, 131
    light in, 33, 38, 134
    love as matrix of, 249

observing orderly motions of, 12
origin of, 18–19, 30
preferred shapes of, 202
quilts charting of the, 222
scientific understandings of, 228
stars in, 41 (*see also* stars)
stories connecting us to, 47–51
through lens of other cultures, 52
wonder and mystery of, 29–30
Crab Nebula, 43, 44, 160
crafts as art, 17
Crane, Stephen, 238
creation stories, 30–32
critical density of universe, 245, 247
Cry of the Earth, Cry of the Poor (Boff), 82n

D

Dahomey, 221
Dan people, 14, 180
Danse Macabre (Fuller), 158
Dante Alighieri, 201
dark energy, 236, 243–244, 246, 247
dark matter, 46, 106, 215–220, 246
darkness, stories connecting us to, 44–46
"dark room" experiment, 55, 57
The Death of Cleopatra (Lewis), 157
depth perception, 33
De revolutionibus orbium coelestium (On the Revolution of the Heavenly Spheres) (Copernicus), 202
dichotomies, 4
dimensions, 163
Dinkinesh, 47, 51
direct imaging, 109
dissonance in art, 240
diversity, 194–195
The Divine Comedy (Dante), 201
"Divine Mind," 205
The Diving Boy (Savage), 156, 158
Dobard, Raymond, 222
Dogon people, 14
Doppler effect, 101–102, 113, 234
Dostoevsky, Fyodor, 105, 130
Douglas, Aaron, 177–184
Douglas, Emory, 229
Douglass, Frederick, 22
doxa, 15
Drake, Frank, 122, 123
Drake equation, 123–125
"Dreams" (Hughes), 106–107
Dresden Codex, 34–37
Du Bois, W. E. B., 182
Dunham, Katherine, x
dwarf galaxies, 211

E

Earth
    axis of, 6, 12
    composition of, 86n
    distance from the Sun, 35, 38
    in heliocentric model, 202
    orbit of, 189, 191, 204
    outer planets' protection of, 189
The Eclipse (Thomas), 89–91
eclipses, 98, 231–232
Eddington, Arthur, 231–232
Egypt
    ancient, 16, 31, 39, 84, 172–174,
        180–181
    Giza, 4–10
Eiffel Tower, 6
Einstein, Albert, 21, 23, 105,
        129–130, 134, 159, 161–163,
        177, 203, 230–233, 235, 236,
        243
electromagnetic radiation/waves,
        57–59
electromagnetic spectrum, 59–62
Eliot, T. S., 197
El Saadawi, Narwal, 3
Emerson, Ralph Waldo, 80, 103
emitted light, 64. *see also* light
Empedocles, 54
energy, light as, 57. *see also* light
Enlightenment, 15, 17, 18
epicycles, 200, 202
episteme, 15
Eratosthenes, 96
Errai, 7
Ethiopia (Fuller), 238, 239
Euclid, 54
Eurocentric worldview, 52
Europa, 206
event horizon, 140
Event Horizon Telescope (EHT),
    144–146
Evers, Medgar, 170
exoplanets, 107
    atmosphere of, 114–115
    belief in intelligent life on,
        121–126
    discovery of, 107–116
    "mini-Neptunes," 116
    Proxima b, 117–120
    "super-Earths," 116
Exoplanet Zoo (Vargic), 109–111
external light, 54, 55. *see also* light
extraterrestrial life (ET), 105–131
    and exoplanet discovery, 107–116
    human desire to discover,
        126–131
    intelligent, 121–126
    on Proxima b, 117–120

F

faith, 227–251
    creative, 236–242
    and island universes, 228,
        230–236
    rational and irrational, 237
    religious, xi
    in what senses do not tell,
        243–251
Falling Star (Bearden), 74, 75
Fermi space telescope, 138
Feynman, Richard, 29
Field Museum, 41
51 Pegasi b, 108–109, 113
fine art, 15, 17
Fitzgerald, Ella, 89
flat universe, 245, 247, 248
"Follow the Drinking Gourd," 222
Fomalhaut, 14
Ford, Kent, 214–216
Frail, Dale, 107–108

Friedmann, Alexander, 233, 234
Fromm, Erich, 237
Fuller, Meta Vaux Warrick, 158,
    238–240, 242
future, stories connecting us to,
    51–52

G

Gaia mission, 210–213
Galactic Disk, 210
galaxies, 47, 49, 50. *see also individual*
    galaxies
    black holes at center of, 46
    dark matter in, 46, 106, 217
    distances to, 235, 236
    dwarf, 211
    and expansion of universe, 49–51
    intelligent life in (*see*
        extraterrestrial life [ET])
    in SMACS 0723, 217–218
    speeds of, 217
    stars in, 46
Galaxy. *see also* Milky Way
    black hole at center of, 44–46
    dark matter in, 46
    halo of, 210
    names given to, 40
    planets in, 115
    star death in, 41, 44
Galeano, Eduardo, 29
Galileo Galilei, 57, 88, 206–209, 225
Gamow, George, 236
Ganymede, 206
gas giants, 112, 113, 206. *see also*
    individual planets

Geller, Margaret, xi
Genzel, Reinhard, 144
geocentric universe, 198, 200, 202
geometry
    in models of the universe,
        244–245
    of pyramids, 7
    of spacetime, 232
Ghez, Andrea, 144
Gibran, Khalil, 175–176
Giza, Egypt, 4–10
globular clusters, 210
gnosis, 15
God
    catching "mind of," 18–19
    in creation story, 30–31
    desire to know, xii
    "Divine Mind" of, 205
    questions about, ix–x
gods and goddesses, 16–17, 34
*God's Trombones: Seven Negro Sermons
    in Verse* (Johnson and Douglas),
    182
Goethe, Johann Wolfgang von, 53,
    68–72, 203, 205
gold headdress ornament (Calima),
    84
Goodson, Albert A., 249
gravitational lensing, 217
gravitational waves, 135–138
gravity, 46
    and black holes, 133–134, 140
        (*see also* black holes)
    in general relativity, 231–232
    nature of, 249
    Newton's law of, 230, 231
    of the Sun, 96
Great Pyramid, 4–7
Green, Jonathan, 71, 72

H

habitable zone, 116
Halley, Edmond, 191
Hall of Negro Life, 177–178, 180
Hampton, Fred, 170
Hanford Observatory, 135
Harlem Artists Guild, 182
Harlem Renaissance, 158, 178,
    180–182, 238
Harlem Woman (Catlett), 148, 149
*Harmonices Mundi (The Harmony of
    the World)* (Kepler), 186–188,
    205
harmony(-ies), 177–195
    color harmony/harmony in color
        theory, 73, 74, 180
    and diversity, 194–195
    and life as music, 184–193

harmony(-ies) *(continued)*
  in nature, 184, 185, 189
  and the newborn universe, 193–195
The Harmony of the World: A Realization for the Ear of Johannes Kepler's, 188
Hawking, Stephen, 18
Heisenberg, Werner, 19–20
heliocentrism, 200, 202–205
helioseismology, 101
Herschel, Caroline, 206, 208
Herschel, William, 40, 206, 209
Hidden in Plain View (Tobin and Dobard), 222
High-Z Supernova Search Team, 243
Hokusai, Katsushika, 152–153, 155
Hollywood Africans (Basquiat), 242
Horn Players (Basquiat), 242
"hot Jupiters," 109
Hubble, Edwin, 49, 235, 236
Hubble Space Telescope, 43, 47, 48, 50
Hubble Ultra Deep Field, 50
Hughes, Langston, 102, 106–107, 182
Human Genome Project, 105–106
humility, 197–198, 205, 220–221
Hunter, Clementine, 22, 238, 239, 241, 242
Huygens, Christiaan, 63, 206
Hyades, 41
hydrogen atoms, 51
Hypatia, 4

I

"I am the Negro woman" (Catlett), 150
Ibn al-Haytham, 54–57, 66
Igbo people, 221
"I have always worked hard in America" (Catlett), 150
"I have given the world my songs" (Catlett), 150
Imara, Nia (art), 83, 162, 241
Improvisation in art, 224–225
Improvisation No. 30 (Cannons) (Kandinsky), 127
"In Harriet Tubman I Helped Hundreds to Freedom" (Catlett), 150, 153, 154
"In Phillis Wheatley I proved intellectual equality in the midst of slavery" (Catlett), 150, 153

"In Sojourner Truth I Fought for the Rights of Women as Well as Negroes" (Catlett), 151, 152
INTEGRAL, 138
intelligent extraterrestrial life, 121–126
intensity, color, 180
interferometry, 145
intermediate-mass black hole binaries, 139
interstellar dust, 209, 211
interstellar gas, 211
interstellar space, 38, 101. *see also* space
intuition, 237
Io, 206
Irony of a Negro Policeman (Basquiat), 242

J

James Webb Space Telescope, 217, 218
Janssen, Jules, 99
Jantjes, Gavin, 32
jazz, ix, 22, 86, 181, 224
Jemison, Mae, 3, 133, 148
Johns, Jasper, xi
Johnson, James Weldon, 182
Johnson, Mark, 20
The Judgment of Paris (Raimondi), 169
Jupiter, 33
  asteroids and comets deflected by, 189
  composition of, 86n
  in heliocentric model, 202
  moons of, 206
  in night sky, 35
  and Sun's wobble, 112, 113

K

Kandinsky, Wassily, 91, 126–130
Kant, Immanuel, 235
Keller, Helen, 227
Kennedy, John F., 170
Kennedy, Robert, 170
Kepler, Johannes, 57n, 115, 177, 186–189, 191, 204–205, 214, 225
Kepler Mission, 115, 116, 123
Kepler space telescope, 114
Khoisan, 31
Khufu, Pharaoh, 4
King, Martin Luther, Jr., x, 170
Klee, Paul, 176
Klimt, Gustav, 165

Kongo cosmogram, 221
Kuba, 221
Kukulkan, 34

L

Lakoff, George, 20
Large Magellanic Cloud, 211–213
Laser Interferometer Gravitational-Wave Observatory (LIGO), 135–139, 146
Laser Interferometer Space Antenna (LISA), 146
Leavitt, Henrietta, 235
Le Concert Champêtre (The Pastoral Concert) (Titian), 167
Le Déjeuner sur l'herbe (Manet), 167, 168
Le Déjeuner sur l'herbe: Les Trois Femmes Noir (Thomas), 167
Lemaître, Georges, 233–234
Leonardo da Vinci, 57
Leonids, 224
Let My People Go (Douglas), 183
Lewis, Edmonia, 157, 158
The Libyan Sibyl (Michelangelo), 181
Lidai mingchen zouyi, 42
life
  building blocks of, 41
  defining, 106
  elements essential for, 44
  extraterrestrial (*see* extraterrestrial life)
  rhythm inherent to, 86
light
  and bending of spacetime, 231–233
  and black holes, 135, 139
  and color, 57–80 (*see also* color[s])
  cosmic microwave background, 247–251
  definition of, 57
  distances traveled by, 38–39
  in early universe, 248
  as electromagnetic spectrum, 59–62
  event horizon for, 140
  from galaxies, 47, 49
  interaction of matter and, 63–66
  memory of, 176
  from neutron stars, 138
  from pulsars, 108
  rainbows, 65
  speed of, 57–59, 163
  from spiral nebulae, 234
  from the Sun, 35, 38, 39, 59–62, 64, 66
  symbolism of, 182, 184

and theories of vision, 54–55
as truth, 54, 67
as universe's storyteller, 33–37
   (*see also* stories)
visible, 59–62
as wave and particle, 57, 59
wavelength, and frequency of,
   57–59
light pollution, 30
Little Dipper, 41
Livingston Observatory, 135
Local Group, 47
Lockyer, Norman, 99

M

Ma'at, 30
Madonna (Catlett), 148
Magellan, Ferdinand, 211
Magritte, René, 105
Malcolm X, 170
Mandala of the Sun God Surya, 85
Man Eating His Heart (Fuller), 238
Manet, Édouard, 167, 168
map quilts, 197–198, 222–223
maps of the sky, 40–41, 210–213
Marius, Simon, 206n
Mars, 33
   composition of, 86n
   in Dresden Codex, 34
   in night sky, 35
   retrograde motion of, 200, 202
Marshall, Kerry James, 168, 170, 171
Martin, Effie Mae, 225
masks, 14–15, 91–94, 148, 180
materialism, 128–130
mathematic arts, 17
Matisse, Henri, 238
matter
   at birth of universe, 194
   in composition of the universe,
      246
   dark, 46, 106, 215–220, 246
   distribution of, 230–231
   interaction of light and, 63–66
Maxwell, James Clerk, 161
Maya, 34–37, 172
Mayor, Michel, 108, 109
McNair, Denise, 170, 171
medicine wheels and sacred hoops,
   13–14
MeerKAT radio telescope, 141–142
M87, 145–146
Mercury, 33
   composition of, 86n
   in heliocentric model, 202
   orbit of, 108, 204, 228, 230
   and Sun's wobble, 113

Merton, Thomas, 171
metaphors, 23, 67, 102–103, 165,
   184, 188–189
Metaphors We Live By (Lakoff and
   Johnson), 20
meteors, 224
Michelangelo, 181
Middle Ages, 17
Milky Way, 39–40, 206–220. *see also*
   Galaxy
   black holes in, 44–46
   center of, 141–143
   composition of, 206
   dark matter in, 217
   different cultures' names for
      God's Backbone, 40
      Hay Merchants Way, 40
      River of Heaven, 40
      Silver River, 40
      Sky's Spine, 40
      Way of Birds, 40
   globular clusters in, 210
   Herschels' map of, 206, 208
   in night sky, 35
   size of, 47
   star births in, 41
   stardust in, 40
   Sun's orbit in, 86
Millennium Simulation Project,
   219, 220
"mind of God," 18–19
"mini-Neptunes," 116
Minkowski, Hermann, 230
Monk, Thelonious, 89
Moon
   ancestors' use of, 30
   in Dresden Codex, 34
   Galileo's drawings of, 208
   light from, 38
   mountains and valleys on, 206
   in night sky, 35
moons, of gas giants, 206
Morrison, Toni, 53, 79–80, 227–228,
   249
Motecuhzoma II, 85
Mother and Child (Picasso), 70, 72
Mother and Child sculptures
   (Catlett), 153, 156
motion
   of the planets, 200, 202, 204
   time and, 161, 163
Munsell, Albert, 72–73
music
   harmony, 86, 178, 181, 182
   hidden meaning in Negro
      spirituals, 222
   life as, 184–193

music of the spheres, 186, 205
   rhythm, 89, 91
"My right is a future of equality with
   other Americans" (Catlett),
   150

N

Nabta Playa, 12–13
NASA, 217
National Endowment for the Arts,
   21
natural philosophy, 17
nature
   art of learning to see, 198–220
      (*see also* conceptions of the
      cosmos)
   humans' relationship with, 19
   as a machine, 19
   Tolstoy's short pieces about, 94
negative space in art, 152
The Negro Woman series (Catlett),
   150–152
Neptune, 34, 86n, 188
Net Casting (Green), 71, 72
neutron stars, 138, 139
Newton, Isaac, 57, 62–63, 66, 68, 69,
   214, 230, 231
NGC 7727, 146
Nisaba, 16, 17
North Star, 6, 7, 180–181
Nwantantay, 92–94. *see also* Bwa
   people

O

objectivism, myth of, 20
objectivity, 19–21
observation of reality, 28
observatories, 12–13, 88, 203–204.
   see also specific observatories
ocher, 10, 11
Odyssey series (Bearden), 76, 79
open universe, 245
Opticks (Newton), 57
Orion Arm, 39, 210, 211
Orion's belt, 12, 39
Orunmila, 17
Overview of the Divine Comedy
   (Caetani), 201

P

paint(s)
   colors of, 69 (*see also* color[s])
   created from peanuts, 27
   oldest form of, 11
Pascal, Blaise, 227

Payne-Gaposchkin, Cecilia, 96
Pecola Breedlove, 227, 249
Pegasus constellation, 108
Penzias, Arno, 247
Perlmutter, Saul, 243
Perseus Arm, 210
philosophy, models of the universe
    and, 203, 205
photons, 59, 63
photosphere, Sun, 97–99
physics, x, 24
    astrophysics, 24, 41, 194
    and black holes, 46, 145
    and expanding universe, 235
    on forces of nature, 249
    Newtonian, 231
    quantum, 19–20, 28
    and sound, 193
    time in, 171
Picasso, Pablo, 70, 72
Pictorial Quilt (Powers), 223, 224
Pilate (Bearden), 78, 79
pinhole camera, 57
planets, 33–34
    Galileo's observations of, 207
    habitable zone for, 116
    movements of the, 200, 202,
        204–205
    "music/sounds of," 186–193, 205
    in night sky, 33
    in our Solar System (*see
        individual planets*)
    outside our Solar System (*see
        exoplanets*)
Plantation (Scott), 198, 199
Plato, 54
Pleiades, 41
Pluto, 188
Polaris, 7
Porter, Cole, 20
positive space in art, 152
Powers, Harriet, 222–224
precession, 6, 12
primary colors and cultural context,
    73
The Prophet (Gibran), 175–176
Proust, Marcel, 165
Proxima b, 117–120
Proxima Centauri, 38, 117, 120
PSR B1257+12, 108
Ptolemy, 54, 198, 200, 201, 203, 225
pulsars, 108
Pumbaa (Savage), 156, 158
pyramids, 4–10
Pythagoras, 185–187, 189

Q

quantum physics, 19–20, 28

Queloz, Didier, 108, 109
quilts, 197–199, 221–225

R

Ra, 30
racism, 227–228
    Basquiat's exposure of, 242
    challenging stereotypes of, 180
    and Douglas's art, 184
    Fuller's exposure to, 238
    and interpretation of color black,
        79–80
radial velocity method, 109, 112–113
Raimondi, Marcantonio, 169
rainbows, 65
reality, 18–20, 28, 160
Realization (Savage), 156
red dwarfs, 120
redshift, 234, 235, 248
reflected light, 63, 64, 66. *see also*
    light
reflecting telescope, 206, 208
refracted light, 63, 66. *see also* light
Reiss, Winold, 182
relativity
    general theory of, 203, 230–233,
        243
    special theory of, 163–165, 230
religions/religious institutions. *see
    also* God; gods and goddesses;
        spirituality
    African art in context of, 14
    in ancient Egypt, 13, 17
    and art, 14, 17, 182
    creation stories, 30–32
    humility in, 205
    Kepler influenced by, 186, 205
    and models of the universe, 202,
        203, 205
    and other life in the cosmos, 121
    religious faith, xi
    and science, 4, 15
    stone circles of, 13
    and the Sun, 86, 186
    universities established by, 17
representational art, 11–14
retrograde motion, 200, 202
Revelations (Ailey), 86
Rhind (Ahmes) Papyrus, 8–9
rhythms, 86–91
ring shout, 82–83, 86
Robertson, Carole, 170, 171
Rodin, Auguste, 238
Rogers, John, 188
Roman, Nancy Grace, 62
rotation curves, 214–216
Rothko, Mark, 91
RR Lyrae stars, 210

Rubin, Vera, 197, 211, 214–216,
    220, 225
Ruff, Willie, 188

S

Sagan, Carl, 123
Sagittarius A*, 44–46, 141, 144–145,
    147
Sagittarius Arm, 210
Sai Baba of Shirdi, 106
Saraswati, 30
Saturn, 33, 190
    composition of, 86n
    in heliocentric model, 202
    inner planets shielded by, 189
    moons of, 206
    in night sky, 35
Savage, Augusta, 156, 158
scattered light, 66. *see also* light
Schmidt, Brian, 243
Schwabe, Samuel, 88
science. *see also individual sciences*
    contextual pursuits in, 15
    cultural roots of, 52
    current practice of, 52
    current systems of, 15
    defining, 21–26
    as essential to human experience,
        10
    European ideas about, 14
    faith in process of, 236–237
    fields of, 17
    human genome mapping,
        105–107
    intersection of art and, x–xii,
        3–28
    making the invisible visible in,
        146, 148
    quantum physics, 19–20
    separation of art and, 3–4
    spirituality and, 128
    systems of, 26
    Tolstoy's short pieces about, 94
    in Western culture, 17–18
Scott, Elizabeth Talford, 197–199,
    224
Seattle, Chief, 52
Seattle Art Museum, 14
Seshat, 16, 17
SETI (Search for Extraterrestrial
    Intelligence), 121–123
    SETI Institute, 121, 122, 125
Shapley, Harlow, 210
Shout (Imara), 83
Shu, 31
Simone, Nina, 22, 67
singularities, 140
Sirius, 14, 39

Skidi Pawnee people, 40–41
sky, maps of the, 40–41
Slave Culture (Stuckey), 82
Slipher, Vesto, 234–235
Slonimsky, Nicolas, 86n
SMACS 0723, 217–218
Small Magellanic Cloud, 211–213
Socrates, 15
solar eclipses, 231–232
Solar System, 34
    closest stars to, 38
    location of, 39, 209, 210
    orbit of the, 86
    planets outside (*see* exoplanets)
solar systems
    diversity of, 115
    multi-planet, 109
solar wind, 99
Solomon, King of Israel, 17
Solovyou, Vladimir, xi
Song of Solomon (Morrison), 79–80
sound. *see also* harmony(-ies)
    of black holes, 137
    "echo" of Big Bang, 247
    listening for alien signals,
        122–123
    and the newborn universe,
        193–195
    sound of silence, 193
    from the Sun, 81, 99–102, 187
Souvenir I (Marshall), 170, 171
space, 133–158
    in art, 146, 148–158
    black holes in, 135–147
    in Catlett's works, 152
    essential "fabric" of, 230
    expansion of, 51
    and flattened perspective,
        152–153
    relative sizes and positions of
        objects in, 153
    and ripples in spacetime,
        134–139
    in theory of special relativity,
        163–164
    and time, 163–164
space telescopes, 43, 47–50, 114, 138,
    217, 218
spacetime
    black holes in, 135–147 (*see also*
        black holes)
    curvature of, 140, 230–231
    dimensions of, 163
    fabric of, 230
    geometry of, 232
    gravitational waves in, 135–137
    ripples in, 134–139
space weather, 89

Special Houses (Catlett), 153
spiral nebulae, 234–235
spirituality, 4. *see also* religions/
    religious institutions
    in art, 23, 126–129, 128, 240
    of Mayan ceremonies, 34
    in nature, 205
    and systems of art and science,
        15, 20–21, 27
stardust, 40
stars. *see also specific stars*
    alignment with Great Pyramid
        shafts, 7
    ancestors' use of, 30
    and ancient astronomical
        creations, 12–14
    birth and death of, 41
    closest to Solar System, 38
    in conceptions of the cosmos,
        210–215
    in creation stories, 31, 32
    cycles of, 12
    death of, 41
    deflection of light from, 231–233
    distance of, 203
    in extrasolar systems, 115
    in galaxies, 46
    light from, 34, 38–39
    maps of, 40–41
    in the Milky Way, 39 (*see also*
        Milky Way)
    neutron, 138, 139
    North Star, 6, 7, 180–181
    patterns of, 33
    pulsars, 108
    quilts depicting, 197–199
    speeds of, 214
    stone circles aligned with, 12–13
    stories connecting us to, 41–44
    Sun-like, 108, 115
    supernovas, 42, 43, 46
    variable, 210
    wobble of, 112, 113
Stela of Aafenmut, 84
Stella, Frank, 133
stellar-mass black holes, 138–144
stellar nurseries, 41
stone circles/arrangements, 12–14
Stone of the Five Suns, 85
stories, 29–52
    connecting us to darkness,
        44–46
    connecting us to the cosmos,
        47–51
    connecting us to the future,
        51–52
    connecting us to the past, 35,
        38–41

connecting us to the stars, 41–44
creation stories, 30–32
light in, 33–37
meaning from, 31, 33
in searching for answers, 30
Stowe, Harriet Beecher, 181
Stuckey, Sterling, 82
subjectivity, 20–21
Sumeria, 16, 17
Sun, 81–103
    ancestors' use of, 30
    and ancient astronomical
        creations, 12, 14
    in art, 84–86, 89–95
    aura of the, 95–99
    barycenter of, 112, 113
    composition of, 96, 98
    and curvature of space, 231–233
    distances of planets from,
        186–191
    gravitational pull of, 214
    habitable zone of, 116
    in heliocentric model, 200,
        202–205
    layers of the, 97–99
    light from, 35, 38, 39, 59–62,
        64, 66
    magnetic fields of, 88–89, 99
    mass of, 41
    in the Milky Way, 209
    orbit of the, 86
    pattern of, 33
    rhythms of, 86–89, 91
    rituals associated with, 81–86
    rotation of, 86, 88
    size of, 96
    in Solar System, 34
    sound from, 81, 99–102, 187
Sun and Plum Branches (Zeshin), 85
sundial, 174
Sun-like stars, 108, 115
sunquakes, 101
Sun Ra, 81
"The Sun's Heat" (Tolstoy), 94–95
"Sun Song" (Hughes), 102
sunspots, 87, 88
Suntzeff, Nicholas, 243
"super-Earths," 116
supermassive black holes, 138–139,
    141, 144–147
Supernova Cosmology Project, 243
supernovas, 42, 43, 46, 138, 243
"Superstition" (Wonder), 89
Surya, 85
Swift, Jonathan, 146
symbolism, 171, 173, 182, 184, 221

T

Tanner, Henry Ossawa, 25, 238
Tarter, Jill, 121, 123, 125, 126, 129
Taurus constellation, 42
techne, 15
Tefnut, 31
telescopes, 145, 205–209. *see also*
    observatories; space telescopes
Tennyson, Lord Alfred, 28
Tesla, Nikola, 53
Theory of Colours (Goethe), 69–70
Thomas, Alma Woodsey, 89–91
Thomas, Mickalene, 167, 168
Thoth, 17
The Three Ages of Woman (Klimt), 165
Thuban, 6, 7
time, 159–176
    essential "fabric" of, 230
    as measure of change, 159–160
    of orbits, 204–205
    as persistent illusion, 161–164
    relative nature of, 163
    remembering, 164–171
    and ripples in spacetime,
        134–139
    and space, 163–164 (*see also*
        spacetime)
    systems of timekeeping, 171–176
Titian, 159, 165–167
Tobin, Jacqueline, 222
Tolstoy, Leo, xi, 24, 94–95
Tompkins, Rosie Lee, 224–225
TON 618, 144
"To S.M., a Young African Painter
    on Seeing His World"
    (Wheatley), 153
transit method, 109, 112–115
transmitted light, 63, 64. *see also* light
TRAPPIST-1, 115, 116
truth, as light, 54, 67
Truth, Sojourner, 150, 181
Tubman, Harriet, 150
Turner, Mary, 238
Two-Part Inventions (Bach), 89

U

ultramassive black holes, 144
uncertainty principle, 19
Underground Railroad, 222
United States. *see also* Western
    culture
    solar storm panic in, 89
    valuation of art vs. science in, 21
universe
    as both static and infinite, 232
    clockwork, 18
    creation of the, 51, 193–194

creation stories about, 30–31
curvature of, 134
expanding, 49–51, 235–236,
    243–244, 248
geometric models of, 244–245
humanity's role in, 19
intimate connection to, 20
models of evolution of, 244
patterns in, 33
size of, 206
as spacetime, 134
static, 228, 230
trying to understand the, 28
    (*see also* conceptions of the
        cosmos)
Untitled ("Black Circe") (Bearden), 77
Untitled (Jantjes), 32
Uraniborg observatory, 204
Uranus, 33–34, 86n, 188

V

van Gogh, Vincent, 64, 67, 159, 160
Vargic, Martin, 109–111
variable stars, 210
Venus, 33, 34
    composition of, 86n
    cycles of, 191, 192
    in heliocentric model, 202
    light from, 38
    phases of, 206
Vera C. Rubin Observatory, 220
very long baseline interferometry
    (VLBI), 145
Virgo (observatory), 138
Virgo Cluster, 47, 49
visible light, 59–62. *see also* light
vision, 54, 55, 59–63
Voltaire, 237

W

Walker, Kara, 152
Washington, Booker T., 26
*The Waterfall Where Yoshitsune Washed
    His Horse at Yoshino in Yamato
    Province* (Hokusai), 152–153,
    155
Wathaurong people, 12
Weiss, Rainer, 135
Wesley, Cynthia, 170–171
Western culture
    art and science in, 17–18
    belief in static universe in, 228,
        230
    black women in art of, 150, 152
    conceptions of the universe in,
        198–205
    creation story of, 31

dichotomies in, 4, 19–21
disconnection in, 129–130
Eurocentric worldview in, 52
ideas of quantum physics in, 20
perception of time in, 159–160
racist interpretations of black in,
    79–80
reframing narratives of, 184
Wheatley, Phillis, 150, 153
Wheeler, John, 140, 244
When Frustration Threatens Desire
    (Marshall), 171
white dwarfs, 243
Wilson, Robert, 247
Wolszczan, Aleksander, 107–108
Wonder, Stevie, 62, 89
Wright, Frank Lloyd, 133
Wurdi Youang, 12

Y

yellow dwarves, 120
Yoruba people, 17, 173
Young, Nettie Lee, 224

Z

Zeshin, Shibata, 85
Zulu series (Jantjes), 32
Zwicky, Fritz, 217

About the Author

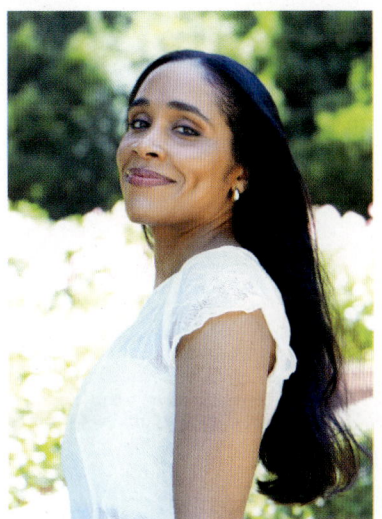

DR. NIA IMARA is an artist and astrophysicist born in Oakland, California. She is a graduate of Kenyon College, received her PhD in astrophysics from University of California Berkeley, did her postdoctoral work at Harvard University, and is currently a professor of astronomy at University of California, Santa Cruz. Her nonprofit organization, Onaketa, provides free STEM tutoring and other educational resources for black and brown youth in the United States and Ghana.